Agricultural Incentives: Past Trends and Prospective Paths towards Sustainable Development

Agricultural Incentives: Past Trends and Prospective Paths towards Sustainable Development

Shashi Shekhar Prasad

RANDOM

RANDOM PUBLICATIONS
NEW DELHI (INDIA)

Agricultural incentives: Past Trends and Prospective Paths towards Sustainable Development

ISBN 978-93-5111-766-7

Published in 2016 in India by

RANDOM PUBLICATIONS

4376-A/4B, Gali Murari Lal, Ansari Road
New Delhi-110 002
Phone : +9111-43580356, 011-23289044, 011-43142548
e-mail: sales@randompublications.com,
info@randompublications.com, randomexports@gmail.com

Reprinted 2019

Type Setting by : Friends Media, Delhi-110089
Digitally Printed at : Replika Press Pvt. Ltd.

Preface

The impact of sector-specific and economy wide policies on agricultural incentives for eighteen developing countries for the period 1975-84 is estimated. The direct effect is measured by the proportional difference between the producer price and the border price. The indirect effect has two components. The first is the impact of the unsustainable portion of the current account deficit and of industrial protection policies on the real exchange rate and thus on the price of agricultural commodities relative to nonagricultural nontradables. The second is the impact of industrial protection policies on the relative price of agricultural commodities to that of nonagricultural tradable goods. We find that in almost all cases the direct effect is equivalent to a tax on exportable goods (–11 percent on average) and to a subsidy for importable (20 percent on average); the indirect effect also taxes agriculture (–27 percent on average) and dominates the direct effect (whether the direct effect is positive or negative); and the direct policies for both importable and exportable stabilize domestic producer prices.

– ***Author***

Contents

1

Farm Business Management: Relation of Rural Economy

A farm's management unit consists of the individual or group responsible for decisions about how a farm will be operated. How a farm is legally organized is often viewed as being the same as its management. A proprietor makes decisions for proprietorships, partners for partnerships, and elected directors and officers for corporate farms.

However, a farm's management unit may not be synonymous with its ownership. For example, land owners may or may not participate in management decisions. The Census of Agriculture reported in 1999 that 14 percent of landlords either made or shared in decisions related to selection of fertilizer and chemicals, while 13 percent helped decide cultivation practices.

FARM MANAGEMENT

Farm management deals with the organisation & operation of a farm with the objective of maximizing profits from the farm business on a continuing basis. The farmer needs to adjust his farm organisation from year to year to keep abreast of changes in methods, price variability & resources available to him. Thus farm management is the science which deals with the analysis of the farming resources, alternatives, choices & opportunities within the framework of resource restrictions & social & personal constraints of farming business. This complex information is integrated and synthesized to increase profitability of the farming business, the ultimate aim being to raise the standard of living of the farming people.

This does not mean that farm management deals exclusively with the maximization of income; in fact, it takes into account the goals and objectives of the individual farmer, other than income maximization. Thus this discipline deals with people or organisers and decision-makers in respect of farms and agricultural production. It is people-oriented rather than crops or livestock per se.

Farm management is a decision-making science. It helps to decide about the basic course of action of the farming business. The basic decisions of the farming business are:

- What to produce or what combination of different enterprises to follow?
- How much to produce and what is the most profitable level of production?
- What should be the size of an individual enterprise, which, in turn, will determine the best overall size of the farm business?
- What methods of production (production practices or what type of quality of inputs and their combination) should be used?
- What and where to market?

CHARACTERISTICS OF FARM BUSINESS SYSTEMS

The 31 percent of farms with the least complex farm business system controlled 11 percent of acres operated and generated less than 3 percent of production value in 2003. Over 85 percent of these farms are rural residences and the rest almost entirely intermediate farms (sales below $250,000 and the operator reports farming as his or her major occupation). Over 98 percent had sales of less than $100,000. These farms specialized in production of field crops other than cash grains or soybeans, beef cattle and general livestock. Operator and operator-spouses managed three-fourths of the least complex farms. They had the highest average operator age and the largest share of primary operators over age 65 years (32 percent). Over 80 percent considered their primary occupation to be off-farm and almost 29 percent were retired. This helps explain the 890 hours worked onfarm by the operator, well below the all-farm average of 1,393 hours.

The most complex farms controlled 8 percent of acres and generated 18 percent of value of production in 2003. Farms in this group were mostly commercial. Over 27 percent had over $500,000 or more in sales (versus 3 percent of all farms). The most complex farms had a much larger share of management teams that included multiple persons. These farms were more common in the Northern Crescent, Heartland, Northern Plains, and Eastern Uplands.

They tend to specialize in dairy, poultry, hogs, cash grains, and soybeans. Nearly two-thirds reported hiring other individuals and three-fourths had custom hire assistance. The primary operators in these management units were younger, averaging 49 years, nearly 7 years less than the all-farm average. On average, primary operators reported working over 3,000 hours on their farms in 2003, with spouses and other operator labour adding more than 1,100 hours to the total.

MEANING OF FARM MANAGEMENT

Farm Management is the application of business principles and the scientific principles of agriculture (as discovered by the chemist, the physicist, the agronomist, the animal husbandmand and other specialists) to the business of farming. The farm manager is the one who plans the farm, arranges the cropping scheme, provides the equipment and actually transacts the business and sees to the operation of the farm.

The term is a new one and it is important that students of farming get a clear understanding of its meaning. Because of the close relation of farm management to soils, farm crops, live stock, and other agricultural subjects, people often confuse the terms and gain erroneous ideas of what farm management really is.

Farm management is also related closely to rural economy. There is danger of confusion arising in the consideration of these terms also. Questions which relate to the single farm as a unit or to a group of farms operated under the same organization are matters of farm mangagement. Those which relate to a group of farms operated separately, or to a community, are matters of rural economy. The farmer must meet both classes of questions. To state it differently, those things that have to do with the organization and operation within the individual farm relate to farm management. Those that have to do with the external relations of the individual farm to the group or community, relate to rural economy.

The Relation of Farm Crops to Farm Management

The farmer must grow crops on his land either to sell or to feed to his live stock. This makes it necessary for him to know how to prepare the land, sow the seed, cultivate the crops and harvest and store the product. In gaining this knowledge he studies the questions from the viewpoint of the agronomist and considers only what will be necessary to provide the best conditions for the growing crop.

The farmer must also decide what crops he will grow. One farmer will want to grow corn, another will grow cotton, and still another will grow hay, wheat or potatoes. Usually he will grow several different crops. He will need to decide how much of each one should be grown. In deciding these questions he will study the questions from the viewpoint of the farm manager.

As an agronomist, the farmer is interested only in how to grow the crop. When the crop is mature and harvested, the problem is completed. As a farm manager he still has to settle the question of what to do with the crop. It may be sold. Would it pay better to feed it to live stock? If so, what kind of live stock will give the best returns for the feed used? These are farm management questions.

Relation of Animal Husbandry to Farm Management

Most farmers grow some live stock. They must know the nature of the animals they wish to grow, the kind of food they eat, how to prepare it for them and when to feed them. They should also know how to select the best animals and to what purpose the various classes of animals are best adapted. In answering these questions they study them from the viewpoint of the animal husbandman and cosider only what is best for the animals.

The farmer must decide, however, what classes of stock he will grow. On some farms dairy cows will be best, on others beef cattle and hogs will be a desireable combination.

There are farms on which horses and sheep with some poultry will be preferable. There are also the question of how much live stock the farm will supply feed for, or what to do with the surplus when more is produced than is required for the live stock. In answering these questions they are studied from the view point of farm management.

Relation of Rural Economy to Farm Management

The management of a farm includes the buying and selling of products. It is important that the farmer have a knowledge of markets and of values, and that he know when to buy and sell. Often cooperative marketing offers the best opportunity for disposing of products at a profit. So long as the farmer is concerned only with the business enterprises relating to his own farm, he is within the field of farm management. When the business activities and social welfare of several farms or an entire community are collectively considered, the question becomes one of rural economy. Farm management deals with the problems of the individual farm from the business point of view. Rural economy deals with the problems of a collection of farms or an entire community.

Studying Farm Organization

The student of farming is taught how to grow crops by the agronomist, how to grow live stock by the animal husbandman. He learns how to know insects and combat them from the entomologist, and how to cooperate with his neighbors in selling his crops or his stock, from the rural economist. Finally, from the viewpoint of a farmer, he will need to select a specific farm, make plans for its arrangement into fields, select the kind of stock and crops that it will be best to raise, and determine how much of each should be raised. He must provide the buildings and equipment, estimate the capital required and the labour necessary to operate the farm. The farm is thus organized into business form and its operation is simplified. This art of farm organization is learned from the farm manager.

Meaning Made Clear by Questions. "How shall I grow corn?" is clearly and agronomy question. How to select and grow cattle, is just as clearly a question for the animal husbandman. "Shall I grow corn? or some other crop? or cattle? Having grown corn, shall I sell it? or feed it to cattle? or hogs?" are farm management questions. If questions similar to the above are put to each problem presented there should be no difficulty in distinguishing the viewpoint from which it should be attacked.

Definition of Terms. In the pages following, reference will be made to the terms, "Capital," "Inventory," "Farm Receipts," "Farm Expenditures," "Farm Income," and "Labour Income." An explanation of these terms is given below.

Capital. Economists divide capital into two classes: (1) Fixed capital, including land, buildings and permanent equipment, as teams, implements and live stock for purposes of production; and (2) circulating capital, which includes values in seed, feed, fertilizers, supplies and market stock or crops as well as cash for operating the farm. The term is limited to those forms which disappear in the process of production, reappearing in other materials of similar kind.

There is a tendency on the part of farmers to tie up most of their capital in fixed forms, leaving too little for circulating or operating capital. The Inventory is a detailed list of the farm property with values assigned. An average of the values at the beginning of the year and at the close is considered to be the value of the capital invested in the business. The difference between the opening and closing inventories shows the increase or decrease in the value of the farm property.

Farm Receipts are derived from two sources: (1) direct and (2) indirect. The direct receipts are represented in the cash received from the sales of crops, live stock, live stock products, for labour for others and from miscellaneous sources. The indirect receipts would arise in an increase in the inventory due to permanent improvements or to appreciated values of any of the various forms of property. Values should not be inflated for the purpose of showing a gain in the inventory. Debts cannot be paid on anticipated rises in land values or in values of property unsold.

Farm Expenditures

Items paid for in cash for the support of the farm only can properly be classed as farm expenditures. Items of personal or family expense should not be charged against the farm in determining the profits from farming. A decrease in the inventory is considered as an indirect expenditure. The amount of the decrease should be deducted, with the direct expenditures, from the receipts to show the net farm income. The Farm Income is the difference between receipts and expenses. It is the return for the use of capital and unpaid labour. The farmer and his investment unite in earning the farm income. To learn what is earned by the unpaid labour only, the amount that the value of the investment would earn if placed at interest, must be subtracted from the farm income.

The Labour Income is the amount earned by the farmer alone. To learn what this amount is, the amount earned by the money invested at the local interest rate and the value of the labour of other members of the family at local wages while employed at labour on the farm, must be deducted from the farm income. The balance will be the labour income of the farmer.

APPLICATION OF FARM MANAGEMENT

As against the general contention that farm management is mostly applicable to crop farms, its application to other types of farming business, such as orchards, dairying, poultry, etc. is no less important. Faculty farm management has contributed more to citrus decline than the poor physico-chemical conditions of the soil, the stock-scion incompatibility, insect pests, disease & viruses. Intercropping with exhaustive crops seems to be one of the major reasons for the poor health of the plants. Citrus die-back or decline does not seem to be a disease by itself, but a mere symptomatic expression of physiological imbalance owing to the cumulative effect of several factors. Similarly, if milk production has to be encouraged, it has to be made more profitable than cereal production. One has to demonstrate how the addition of mulch animals to a farm fits into its resource organisation & yields higher returns.

With land as a scarce resource, efforts are made to intensify its use for maximising profits through multiple cropping. During a given period, the relative profitability of different crop rotations is evaluated & finally selected for adoption. 'Kufri Sinduri' variety of potato is not being adopted on a large scale in spite of its giving 10 percent higher yield than some other varieties, *e.g.* 'Up-to-date'.

The reason is simple. The former is a late variety & delays the sowing of wheat which decreases its yield. It is not 'Kufri Sinduri' potatoes versus 'Up-to-date' potatoes; rather, it is 'Kufri-Sinduri' followed by wheat versus 'Up-to-date' followed by wheat that counts in farm management. Similarly, in West Bengal, where jute has been competing with paddy, the situation is changing with the introduction of high-yielding varieties of wheat, which can be rotated with paddy.

The comparison in terms of the relative profitability now runs between jute & paddy plus wheat, & not just between jute & paddy. Owing to the paddy-wheat rotation becoming more profitable, jute production in West Bengal is suffering a set back. As Indian agriculture becomes more productive with the passage of time, the farmers will be confronted with the task of making the best selection from among many alternatives.

SUNDRY FARM BUILDINGS

FARM WORKSHOP FACILITIES

A workshop provides a focal point at the farmstead for the repair and maintenance of machines, implements and structures. It also provides a place where tools can be stored in an orderly manner, a store for supplies and spare parts, and a shelter where work can be carried out during inclement weather. A facility of this type should be available on every farm. The size and design of a workshop, however, should be commensurate with the size of the farm and the work to be done in the shop.

The small holder may be adequately served with a tool storage cupboard that can be locked for security and a workbench with a simple homemade vice for holding tools while they are being sharpened or fitted with new handles. From this simple beginning a more complete facility may gradually evolve as the farm operation grows and more equipment is required. Since repair tools and supplies represent a considerable investment, most farmers will want to store them in a secure place. Many small scale farmers will not require a separate store for this purpose, but if stored together with hand tools and small implements, the number of items may motivate the farmer to build a storeroom by enclosing part of the workshop with solid walls. A simple work shelter and store suitable for repair work and the storage of small implements. Note that the doors to the store may be designed with racks and hooks to hold supplies and tools. Fuels and other combustible materials should not be stored with the tools. A simple work bench and vice can also be housed under the shelter.

At the other extreme, a large ranch or commercial farm may need a separate building with extensive equipment for maintaining the farm machinery, tractors and vehicles. A farmer may also use his workshop to do routine repairs and preventive maintenance during the off season, to build or modify some of the equipment used on the farm and to prefabricate building elements to be used in construction projects.

The workshop facilities should be cost effective. That is, enough savings should be realized from timely maintenance, repairs and construction projects to pay for the cost of the building and the necessary tools and equipment. Although it is difficult to put a monetary value on timeliness, there is no question that being able to make emergency repairs is important. Some farm operations (planting, spraying, milking) are more sensitive than others to prolonged interruptions, and having facilities to complete repairs on the farm can reduce delays to a minimum.

Other factors, apart from the farm size, which will influence the extent of the workshop facilities are the number and diversity of machines, the availability of service from dealers, and the interest and mechanical skill exhibited by the farmer and farm labourers. If necessary, a skilled mechanic may be employed. Without qualified personnel to use the shop it becomes questionable in value and may even contribute to more frequent breakdowns and additional expense due to careless work.

The workshop should be located close to the work centre of the farm and convenient to the farm home on ground that is well drained and sufficiently level to allow easy maneuvering of equipment. Where electric power is available, proximity to the power source should be considered. In tropical climates the workshop may be a simple pole structure with a non-flammable roof. Unless dust is a problem, it may be feasible to leave the sides open to provide good light and ventilation.

Heavy-gauge wire netting can be used to make the area more secure without reducing light or ventilation. A pole structure of this sort can be enclosed with offcuts or corrugated steel at a later time, but if this is done, there must be provision for several good-sized windows.

While a simple earth floor is often satisfactory, concrete offers the advantages of an easily cleaned, level surface. To do a clean repair job, a clean work area is essential and this is particularly important when lubricated mechanisms are reassembled. The level surface is helpful in some assembly or alignment operations.

The following additional features are important for a safe and efficient shop:

- Sufficient room for the largest machine that may need repair, including workspace around it. If the machine is large, truss roof construction may be needed to provide the required space without intermediate supports.
- An entrance that is both wide enough and high enough for the largest equipment that the shop has been designed to accommodate. If the building is enclosed with either solid walls or wire netting, a second door is essential for safety in case of fire.
- Some means of lifting and supporting heavy loads. When the roof span is 3m or less, a timber beam is often adequate. For larger spans or very heavy loads a truss will be required. Alternatively, a portable hoist can be used.
- Electric lighting and electrical service for power tools.
- A water supply for both convenience and safety.
- One or more fire extinguishers of a type suitable for fuel fires. Two or three buckets of dry sand are a possible substitute or supplement for a fire extinguisher.
- Storage cabinets for tools, supplies and spare parts. Sturdy doors can be locked for security and also provide space to hang tools and display small supplies for easy access.

- A heavy workbench attached to the wall or otherwise firmly supported. It should be 1 m high, up to 800mm deep and at least 3m long and equipped with a large vice. There must be sufficient clear space around it to maneuver workpieces and, if attached to a solid wall, ample window openings above it to provide light.

Equipment needed in the workshop will depend on the type and extent of work to be done. Generally this means those tools required to perform day-to-day maintenance on machines and to carry out general repair work and small construction jobs required on farm buildings and equipment. However, any shop, regardless of size, will need some simple woodworking tools, some means of sharpening field tools, and wrenches (spanners) of various types and sizes. If the shop equipment includes a welder, it should be located, in the interest of safety, away from the woodworking area and preferably near the main door where it can conveniently be used inside or outside the building.

Flammable materials such as sawdust, shavings and oily rags must never be allowed to accumulate in the workshop since they represent a fire hazard, and fuels should be stored in a separate area. Generally good order and cleanliness in the shop makes for efficient work, convenience and safety.

MACHINERY AND IMPLEMENT STORAGE

On many small-scale farms in Africa all cultivation and transport operations on the term are done manually. The few small-sized hand tools and implements used in such farming can normally be stored in any multipurpose store at the farmstead. The store needs only to be secure for protection of the equipment from theft and vandalism, and dry so as to avoid deterioration of the metal and wooden parts.

The tools will last longer if they are cleaned and working surfaces are greased prior to storage. The tools may be hung on rails or hooks on the wall or from the ceiling for order and convenience and to protect them from dampness penetrating an earth floor in the store. Implements such as ploughs, harrows and cultivators are damaged little by rust when left outdoors. If they are properly cleaned prior to storage and metal surfaces, particularly all threaded parts used for adjustments, are greased, then a little rust is not likely to harm performance enough to justify the cost of a storage structure. A fenced compound can offer adequate protection against theft during storage. Although implements containing wooden parts are more susceptible to decay, those parts can usually be replaced at low cost. Tractors and other complex machines will function better when needed if they have been stored under cover and given a complete off-season check-up. An adequate storage structure for these machines is likely to be economically feasible.

For most purposes a narrow open-side shed with a welldrained, raised earth or gravel floor will be adequate for machinery storage. The sides of the building can be partly or wholly enclosed with netting or solid walls when security conditions make it necessary. The building must be high enough to accommodate the highest machine. A smooth, level floor makes it easier to attach and detach tractor-mounted equipment or to move other machines. The space required can be determined by obtaining the dimensions of all the machines and implements to be stored. Then, using graph paper, the outline of the machines can be sketched onto a plan view, allowing additional space for maneuvering. Any roof-supporting posts inside the building or in the open sides must be marked on the drawing, since they will restrict the way the floor space can be utilized. Since many machines can not be easily moved, it is desirable to arrange the stored machines so that shifting is unnecessary.

Fire resistant construction is desirable where tractors, cars and other powered machines are stored. A pole structure with an earth floor, sheet metal walls, timber trusses and metal, asbestos-cement or sisal-cement roofing will offer adequate fire resistance. Machinery stores and farm workshops are constructed in much the same way and are usually placed close together for convenience.

In fact, they may be housed in one building with a workshop section at one end and machinery and implement storage in the balance of the building.

FUEL AND CHEMICAL STORAGE

Many materials that are used on farms fall into the category of "hazardous materials", since they are either highly flammable or poisonous. The type and quantities of these materials requiring storage will vary from one farm or one cooperative store to the next and only a few basic requirements for safe storage will be considered here. Other materials frequently used on farms such as fertilizers and cement also have special storage requirements mainly because they are hydroscopic, *i.e.*, they tend to pick up moisture from the atmosphere.

FARM-EFFICIENCY MEASURES

Efficiency is the ratio of the output to the input. This concept is important as it shows how much profitable the farming business is. Various measures to explain the efficiency of the business as a whole, & in parts, are available. When examined together they help to point out the weaknesses in the farm business & provide a guideline as to which part of the business deserves special attention for making improvements. In a particular situation due importance is given to a particular measure.

For instance different measures would be adopted for indicating the volume or size of the business, aggregate earnings from particular factors or the business as a whole & returns per unit of a particular factor input. Further, the efficiency of a farm can be judged from the costs or returns or both. No single efficiency measure is so complete as to give a true picture of the entire farm business.

These efficiency measures help to reorganise the same farm for which they are calculated but they must be used with great caution, while comparing different farms. This is especially important in the developing countries where farming is of diversified nature & individual farm business has wide variations in respect of soil type, resource restraints, capacities & capabilities of the farmers to undertake risks, their attitude towards innovations, etc.

Also, these measures suffer from limitations for making comparisons because of their changing prices, costs, & conditions of the farm business. Adjustments with the price & coat indices are necessary before such comparisons give any valuable information.

FARM CREDIT

Analysis for credit. A banker's plan is something more than the farm management specialist's plan. A banker is more concerned with the recovery of the loan advanced by him to the farmer. A farm-management specialist's plan is all right, but a banker, if he is to call it his plan, must examine it, as not only the additional returns from the alternative plan, but also whether the farmer would be able to repay the loan from the additional returns or not. The credit forms the total capital requirement minus the farmer's owned funds. Most of the banks expect the farmer to contribute 25% of the total capital requirement as margin money so that he has some stake in the business.

Season-wise analysis. Because the loan is to be paid at the end of every season, repayments include:

(a) The total amount in the case of self-liquidating loans which are for short periods & for purchasing seeds, fertilisers, hiring casual labour, etc.

(b) The instalments of the medium & long-term loans which are not self-liquidating & are for constructing buildings, raising irrigation structures, purchasing implements, machinery etc.

INTENSIVE FARMING OF AGRICULTURE

Intensive farming or intensive agriculture is an agricultural production system characterized by the high inputs of capital, labour, or heavy usage of technologies such as pesticides and chemical fertilizers relative to land area. This is in contrast to many forms of sustainable agriculture such as organic farming or extensive agriculture, which involve a relatively low input of materials and labour, relative to the area of land farmed, and which focus on maintaining long-term ecological health of farmland, so that it can be farmed indefinitely.

Modern day forms of intensive crop based agriculture involve the use of mechanical ploughing, chemical fertilizers, herbicides, fungicides, insecticides, plant growth regulators and/or pesticides. It is associated with the increasing use of agricultural mechanization, which have enabled a substantial increase in production, yet have also dramatically increased environmental pollution by increasing erosion, poisoning water with agricultural chemicals, and destroying forests to make room for farmland.

Intensive animal farming practices can involve very large numbers of animals raised on limited land which require large amounts of food, water and medical inputs (required to keep the animals healthy in cramped conditions). Very large or confined indoor intensive livestock operations (particularly descriptive of common US farming practices) are often referred to as Factory farming and are criticised by opponents for the low level of animal welfare standards and associated pollution and health issues.

Advantages

Intensive agriculture has numbers of benefits:

- Significantly increased yield per acre, per person, and per dollar relative to extensive farming.
- Not much space for the animal(s) to move therefore less energy used up; so less food supplied to the cattle, which leads to cheaper products.
- The animals are being fed continuously for maximum growth, and they have very little exercising space to burn off the extra meat.

Disadvantages

Intensive farming alters the environment in many ways.

- Limits or destroys the natural habitat of most wild creatures, and leads to soil erosion.
- Use of fertilizers can alter the biology of rivers and lakes. Some environmentalists attribute the hypoxic zone in the Gulf of Mexico as being encouraged by nitrogen fertilization of the algae bloom.
- Pesticides generally kill useful insects as well as those that destroy crops.
- Generally not sustainable — often results in desertification, or land that is so poisonous and eroded that nothing else will grow there.
- Requires large amounts of energy input to produce, transport, and apply chemical fertilizers/pesticides
- Use of chemicals on fields creates run-off, excess runs off into rivers and lakes causing pollution.
- Use of pesticides have numerous negative health effects in workers

who apply them, people that live nearby the area of application or downstream/downwind from it, and consumers who eat the pesticides which remain on their food.

Pre Modern Intensive Farming

Pre modern intensive farming techniques and structures include terracing, rice paddies, and various forms of aquaculture.

Oysters

Oysters were likely the first sea animal to be transported from one area to another and cultivated as food. The ancient world, while knowing little about the reproduction of oysters, knew much about the conditions necessary for their growth. Pliny the Elder, a noted Roman naturalist of the first century, left an account of artificial oyster beds established in Lake Lucrinus near Naples by a Sergius Orata about 95 B.C. Orata's methods consisted of preparing the grounds by removing other forms of marine life, planting seed oysters, cultivating the oysters by keeping them separated in order to grow to a well-formed, mature size, and finally harvesting them when they were ready for market. Modern oyster farming, based on the knowledge of oyster biology, basically follows the Roman procedure. Fisheries and Oceans Canada article *American Oyster*

Terrace

In agriculture, a terrace is a leveled section of a hilly cultivated area, designed as a method of soil conservation to slow or prevent the rapid surface runoff of irrigation water. Often such land is formed into multiple terraces, giving a stepped appearance.

The human landscapes of rice cultivation in terraces that follow the natural contours of the escarpments like contour ploughing is a classic feature of the island of Bali and the Banaue Rice Terraces in Benguet, Philippines. In Peru, the Inca made use of otherwise unusable slopes by drystone walling to create terraces.

Rice Paddy

A paddy field is a flooded parcel of arable land used for growing rice and other semiaquatic crops. Paddy fields are a typical feature of rice-growing countries of east and southeast Asia including Malaysia, China, Sri Lanka, Myanmar, Thailand, Korea, Japan, Vietnam, Taiwan, Indonesia, India, and the Philippines. They are also found in other rice-growing regions such as Piedmont (Italy), the Camargue (France) and the Artibonite Valley (Haiti). They can occur naturally along rivers or marshes, or can be constructed, even on hillsides, often with much labour and materials.

They require large quantities of water for irrigation, which can be quite complex for a highly developed system of paddy fields. Flooding provides water essential to the growth of the crop.

It also gives an environment favourable to the strain of rice being grown, and is hostile to many species of weeds. As the only draft animal species which is adapted for life in wetlands, the water buffalo is in widespread use in Asian rice paddies.

There are significant adverse environmental impacts from rice paddy cultivation due to the generation of large quantities of methane gas. World methane production due to rice paddies has been estimated in the range of 50 to 100 million tonnes per annum; this level of greenhouse gas generation is a large component of the global warming threat and derives simply from an expanding human population.

Rice-farming and the use of paddies in Korea is ancient. Korean paddy-farming can provide cultural background on the use of paddies in East Asia. A pit-house at the Daecheon-ni site yielded carbonized rice grains and radiocarbon dates indicating that rice cultivation may have begun as early as the Middle Jeulmun Pottery Period (c. 3500-2000 B.C.) in the Korean Peninsula. The earliest rice cultivation in the Korean Peninsula may have used dry-fields instead of paddies.

The earliest Mumun features were usually located in low-lying narrow gulleys that were naturally swampy and fed by the local stream system. Some Mumun paddies in flat areas were made of a series of squares and rectangles separated by bunds approximately 10 cm in height, while terraced paddies consisted of long irregularly shapes that followed natural contours of the land at various levels. Mumun Period rice farmers used all of the elements that are present in today's paddies such terracing, bunds, canals, and small reservoirs. We can grasp some paddy-farming techniques of the Middle Mumun (c. 850-550 B.C.) from the well-preserved wooden tools excavated from archaeological rice paddies at the Majeon-ni Site. However, iron tools for paddy-farming were not introduced until sometime after 200 B.C. The spatial scale of individual paddies, and thus entire paddy-fields, increased with the regular use of iron tools in the Three Kingdoms of Korea Period (c. A.D. 300/400-668).

Modern Intensive Farming Types

Modern intensive farming refers to the industrialized production of animals (livestock, poultry and fish) and crops. The methods deployed are designed to produce the highest output at the lowest cost; usually using economies of scale, modern machinery, modern medicine, and global trade for financing, purchases and sales. The practice is widespread in developed nations, and most of the meat, dairy, eggs, and crops available in supermarkets are produced in this manner.

Sustainable Intensive Farming

Biointensive agriculture focuses on maximizing efficiency: yield per unit area, yield per energy input, yield per water input, etc. Agroforestry combines agriculture and orchard/forestry technologies to create more integrated, diverse, productive, profitable, healthy and sustainable land-use systems. Intercropping can also increase total yields per unit of area or reduce inputs to achieve the same, and thus represents (potentially sustainable) agricultural intensification. Unfortunately, yields of any specific crop often diminish and the change can present new challenges to farmers relying on modern farming equipment which is best suited to monoculture. Vertical farming, a type of intensive crop production that would grow food on a large scale in urban centres, has been proposed as a way to reduce the negative environmental impact of traditional rural agriculture.

Intensive Aquaculture

Aquaculture is the cultivation of the natural produce of water (fish, shellfish, algae, seaweed and other aquatic organisms). Intensive Aquaculture can often involve tanks or other highly controlled systems which are designed to boost production for the available volume or area of water resource.

Intensive Livestock Farming

The modern examples of intensive farming are broadly referred to as Concentrated Animal Feeding Operations (CAFOs) or often termed Factory farming. These include:

- Intensive pig farming or Intensive piggery farming
- Large scale chicken farms
- Cattle feed lots.

Individual Industrial Agriculture Farm

Major challenges and issues faced by individual industrial agriculture farms include:

- Integrated farming systems
- Crop sequencing
- Water use efficiency
- Nutrient audits
- Herbicide resistance
- Financial instruments (such as futures and options)
- Collect and understand own farm information;
- Knowing your products
- Knowing your markets

- Knowing your customers
- Satisfying customer needs
- Securing an acceptable profit margin
- Cost of servicing debt
- Ability to earn and access off-farm income;
- Management of machinery and stewardship investments.

Integrated Farming Systems

An integrated farming system is a progressive biologically integrated sustainable agriculture system such as Integrated Multi-Trophic Aquaculture or Zero waste agriculture whose implementation requires exacting knowledge of the interactions of numerous species and whose benefits include sustainability and increased profitability.

Elements of this integration can include:

- Intentionally introducing flowering plants into agricultural ecosystems to increase pollen-and nectar-resources required by natural enemies of insect pests
- Using crop rotation and cover crops to suppress nematodes in potatoes.

Crop Rotation

Crop rotation or crop sequencing is the practice of growing a series of dissimilar types of crops in the same space in sequential seasons for various benefits such as to avoid the build up of pathogens and pests that often occurs when one species is continuously cropped. Crop rotation also seeks to balance the fertility demands of various crops to avoid excessive depletion of soil nutrients. A traditional component of crop rotation is the replenishment of nitrogen through the use of green manure in sequence with cereals and other crops. It is one component of polyculture. Crop rotation can also improve soil structure and fertility by alternating deep-rooted and shallow-rooted plants.

Water use Efficiency

Crop irrigation accounts for 70% of the world's fresh water use. The agricultural sector of most countries is important both economically and politically, and water subsidies are common. Conservation advocates have urged removal of all subsidies to force farmers to grow more water-efficient crops and adopt less wasteful irrigation techniques.

For crop irrigation and plant irrigation, optimal water efficiency means minimizing losses due to evaporation or runoff. An evaporation pan can be used to determine how much water is required to irrigate the land. Flood irrigation, the oldest and most common type, is often very uneven in distribution, as parts of a field may receive excess water in order to deliver sufficient quantities to other parts. Overhead irrigation, using centre-pivot or lateral-moving sprinklers, gives a much more equal and controlled distribution pattern, but in extremely dry conditions much of the water may evaporate before it reaches the ground. Drip irrigation is the most expensive and least-used type, but offers the best results in delivering water to plant roots with minimal losses.

As changing irrigation systems can be a costly undertaking, conservation efforts often concentrate on maximizing the efficiency of the existing system. This may include chiseling compacted soils, creating furrow dikes to prevent runoff, and using soil moisture and rainfall sensors to optimize irrigation schedules.

Water catchment management measures include recharge pits, which capture rainwater and runoff and use it to recharge ground water supplies. This helps in the formation of ground water wells etc. and eventually reduces soil erosion caused due to running water.

Nutrient Audits

Better nutrient audits allow farmers to spend less money on nutrients and to create less pollution since less nutrient is added to the soil and thus there is less to run off and pollute. Methodologies for assessing soil nutrient balances have been studied and used for farms and entire countries for decades. But at present "there is no standard methodology for calculating nutrient budgets and there are no accepted 'benchmarks' figures against which to assess farm nutrient use efficiency. [A standard methodology] for calculating nutrient budgets on farms diffuse water and air pollution from agriculture [through] best management practices in the use of fertilisers and organic manures, as part of the continued development of economically and environmentally sustainable farming systems."

Herbicide Resistance

In agriculture large scale and systematic weeding is usually required, often performed by machines such as cultivators or liquid herbicide sprayers. Selective herbicides kill specific targets while leaving the desired crop relatively unharmed. Some of these act by interfering with the growth of the weed and are often based on plant hormones. Weed control through herbicide is made more difficult when the weeds become resistant to the herbicide. Solutions include:

- Using cover crops (especially those with allelopathic properties) that out-compete weeds and/or inhibit their regeneration.
- Using a different herbicide

- Using a different crop (*e.g.* genetically altered to be herbicide resistant; which ironically can create herbicide resistant weeds through horizontal gene transfer)
- Using a different variety (*e.g.* locally-adapted variety that resists, tolerates, or even out-competes weeds)
- Ploughing
- Ground cover such as mulch or plastic
- Manual removal.

FARMING MANAGEMENT PRACTICES

Farming management practices include land preparation, planting, thinning and weeding, fertilizer application, pest control, irrigation, harvesting, and post-harvest processing. Each of the processes mayor may not affect the amount of genetic diversity of the crop population over time. Different levels of fertilizer application or the use of organic rather than chemical fertilizer may select for different genotypes in a population. Irrigated and non-irrigated populations of faba beans in Morocco have different population genetic characteristics. Dense planting to reduce weeding and post-harvest storage of seeds also play a selection role on seed survival and the continuation of subsequent traits in the next generation.

Key questions for national on-farm conservation programmes are (1) which farming management practices influence genetic diversity and (2) to what extent do these practices affect the amount of genetic variation in the crop population. For national programmes to look at these questions, interviews and observations of farmer management practices are needed. Field trials using different degrees of density or fertilizer application coupled with measurements of genetic diversity may be desirable to investigate specific effects of particular treatments. Such studies may, however, lie outside the capacity of this particular project and be best explored in specific case studies.

Environmental selection: When a farmer decides to plant a cultivar on a particular micro-site, the crop population is exposed to specific environmental selection. By planting barley in soils prone to water logging or in different agroecological sites, the Ethiopian farmer is subjecting the plants to environmental selection for tolerance to a specific stress. Similarly, by planting upland "Jumla" rice at elevations up to 3000 m in Nepal, or "Chao" rice on shallow soils in northern Vietnam, the farmer is exposing a crop population to cold stress or poor soil conditions.

Questions under this selection category revolve around (1) how environmental selection has influenced the genetic diversity of the population over time and (2) which environmental conditions playa significant role in affecting the amount of genetic variation in the crop population. The farmer's involvement is in making the decision to plant that particular population in a particular micro-habitat. To answer these questions, national programmes will need to focus on basic questions of ecological genetics, to understand how populations adapt to their environments, and to determine the type and extent of effect of environmental factors on the amount of genetic variation over time. The extent to which farmers maintain unique types for particular micro-environments will be important, as well as the occurrence within local cultivars of G × E interactions or broad tolerance of a range of stress environments. Social scientists' involvement will be in investigating what factors influence a farmer to plant in the particular habitat.

The work will likely point to specific studies over longer periods that will help breeders and users understand the nature of adaptation in local cultivars. *Population size:* The size of a population planted by a farmer will affect the amount of genetic variation of the crop population over time. The smaller the population, the more likely it is that genetic drift, inbreeding, loss of alleles, and stochastic events will affect the population. From a conservation perspective, crucial questions include: How does population size influence the genetic diversity of the local cultivar population? What is the effective population size that ensures long-term stability of the population? Should a national programme be concerned with a group of smaller populations or metapopulations, or should the conservation focus be on individual farmer fields? To understand the effect of size on the amount of genetic variation of a population requires minimum population viability studies and population genetic methods.

Seed source

A farmer makes a choice each year on what percentage of his or her own seeds to save and plant and what percentage he or she will acquire from other farmers. How do migration (influx of new seeds) and bottlenecks (reduction in the number of saved seeds) affect the genetic diversity of the population over time? Louette and Smale (1996) have shown that after six crop rotations of the farmer studies in Mexico, only 48% of the seed material remained from the farmer's original stock.

This is the information that a population geneticists will need to determine effective population sizes. Again, the question of whether a metapopulation should be the unit of concern for national conservation programmes is important. This involves social scientists and community-based groups developing an understanding of the seed supply system in order for plant geneticists to be able to determine the effective population size for the cultivar in question. This requires knowledge, not only of seed source and informal seed supply systems, but also of how different storage systems influence the survival of different seeds over time.

On-farm data for conservation and use: On-farm maintenance of local cultivars is currently being promoted as a strategy to conserve genetic diversity and as a process that will secure the continued availability of genetic variation while ensuring the continuing betterment of farmer livelihood over time. In the past, most conservation workers have focused largely on *ex situ* conservation, with the expectation that local cultivars maintained over the centuries will shortly disappear. This is now seen as an oversimplification of the complex issues at play (Louette, this volume). Knowledge of the processes involved in farmer selection and maintenance of crop cultivars can be used by national programmes to make decisions on where to support, assist, or intervene to promote the conservation of genetic diversity and strengthen the effectiveness of conservation. How can this same information, linking farmer decision making to genetic diversity, be used to help national programmes improve the livelihood of the farmer? To be maintained by farmers, crop genetic resources must have value to them and must retain this value and be competitive to other options a farmer might have, or even better, to increase this value such that it is reflected in an increase in a farmer's standard of living.

Value may be added to crop resources in two main ways: the material itself may be improved or the demand for the material or some product may increase. One option is to seek improved quality, disease resistance, yield, taste, or other farmer preferred characteristics, through decentralized breeding activities. Sthapit, Joshi, and Witcombe (1996a) have shown that, by utilizing farmers' knowledge, acceptable varieties can be bred with minimum use of resources in the high altitude areas of Nepal. Effective support for farm based maintenance needs to explore the ways in which this can best be done and the impact it has on the extent and distribution of genetic diversity.

Value can also be added to crop resources by better processing, storage, and marketing, where the farmer receives more benefit from the final product. An important role for community-based organizations and nongovernmental organizations is the formation of farmer cooperatives and farmer-managed community seed banks to maximize returns to the farmers themselve. Government policy may also playa role in ensuring that farmers' inputs and seeds receive the same or better market treatment and government support as improved varieties.

Conclusion: The majority of early studies on on-farm conservation have focused on trying to determine what factors have caused a farmer to maintain or not maintain a diversity of local cultivars. Other studies have looked at the amount of genetic variation of these cultivars at a given moment in time. The research discussed in this chapter aims to take earlier studies further by attempting to quantify over time how specific farmer-based decisions may determine both the amount of genetic variation and the effective population size of a cultivar population over time.

Changes to the amount of genetic variation can be measured as can the effect of farmer-based management on the effective population sizes. Genetic variation may be linked to farmer decisions in five major categories: agromorphological characteristics, farm management practices, planting location, size of the population, and seed source. These decisions in turn are based on environmental and socioeconomic influences. The first link forms the basic data set for the second link. Understanding these processes requires the involvement of people from informal and formal institutes and from a variety of scientific disciplines. Once there is better knowledge of these processes, national programmes will be in a better position to support, assist, or intervene in the conservation of genetic diversity on-farm.

Farmers themselves will ultimately determine the extent to which on-farm maintenance of local cultivars continues and, hence, contributes to the overall conservation of crop genetic diversity. Previous conservation work underplayed or even ignored this contribution and emphasized *ex situ* methods of diversity maintenance. Through the creation of a framework of knowledge of the processes involved, it is hoped to provide a more complete understanding that redresses this imbalance.

THE BUSINESS SIDE OF FARMING

FARMING AS A BUSINESS

Farming has not usually been considered a business. The diversity of the duties of the farmer, the area over which the operations of the farm extend and the complexity of the records required, combine in making difficult the organization of the details of farming into business form. The successful financial operation of a farm presents quite as complex problems and calls for at least as much business ability and judgment as is required in operating a store with the same investment. Farming, therefore, should be considered as a business, and the man who can produce his crops and products at the lowest cost and sell them at the highest price, investing the proceeds to the best advantage, should be considered the best farm manager. The man who knows the details of the cost of production and operation, and whose records show the profitable and unprofitable lines of production, thus enabling him to eliminate those that do not yield a profit, may be counted as the best business man.

A farmer should know the elements of soil fertility. He must understand the principles of the movement of soil water, and the action of soil bacteria. He should understand the nature of plant growth and be familiear with varieties and species of plants and with the effect of one crop on the crop following. He must also be familiar with animals and their habits and know how to feed and care for them. In addition, he must know how to buy and sell to advantage, make contracts, and plan his buildings and his farm so as to necessitate a light expenditure for labour, also that he may distribute his labour to advantage over the various farm enterprises. And he should know how to keep accounts. The farmer in organizing his business could well follow the example of the merchant. The merchant first takes an inventory of his stock. He studies the demand for his goods, both present and prospective. He notes the supply, the cost, and the demand for each article.

He calculates the labour required to operate his business and such other items of expense are considered as may be legitimately charged against the business. He regulates his purchases and his prices according to the cost of securing his goods and putting them on the market. In conducting a large store business, it is customary to organize it into departments, putting some competent person in charge of each department and having the labour and accounting charges so systematized and recorded as to show the profit or loss from each department and from the business as a whole.

The farmer should likewise take an inventory of his capital, stock, and equipment. He should consider the type of farming to which the soil and climate are adapted. He should consider the fertility of the soil and the demand that will be made upon it by the crops grown. He should consider, in connection with the soil fertility, the sources from which it may be renewed and at what cost.

He must study the markets, the transportation, and the demand for such crops as he grows; also the cost of producing each of the crops and the probable net profit that will be returned. His labour likewise should be charged against the various crops or enterprises and distributed to the best advantage. In studying the problems of farm organizations, interest on investment, taxes, insurance, and other expense must be included as they affect the financial result. As in a large store business, it is frequently necessary to organize the large farm into departments, keeping accounts with the dairy, with the swine, the grain crops, the garden, and other similar enterprises. Where the business is large enough, it is well to put an expert in charge of each large branch or group of enterprises, thus enabling one to use cheaper labour for performing the work or making the labour more effective by closer supervision.

Where the farming is conducted as an organized business, and accounts are kept with the various lines of work, it is possible at the end of the year, to make a business statement which will show which lines have been profitable. The manager then can change his methods or drop out those lines that prove to be unprofitable and the business as a whole may be put on a better basis.

Investment

The investment of money in land, buildings, and equipment demands careful consideration. It is possible to pay so much for a farm that it will be impossible to produce sufficient revenue to meet the expense of operation and to pay a normal rate of interest on the money invested. This is particularly true where low-priced products are produced. A farm may be highly productive but so located that it will be impossible to market the produce on a profit bearing basis. One should study closely the market facilities of the neighbourhood and raise supplies which can be successfully marketed locally, or which can be transported to a market that pays well for such produce. Unless the produce of the farm is well related to the market, the farm is likely to be operated at a loss.

Proportion in Real Estate

It is a mistake for one to invest all of his capital in the real estate itself. Sufficient capital should be reserved for operating the farm. The hunger for land has induced many farmers to buy more land that they can equip and operate well. Such farmers are said to be land poor. They would secure greater profit from a medium sized farm, well tilled and managed, than from a large one which is insufficiently equipped and poorly operated. Rarely should more than 50 or 60 per cent of the capital be tied up in the land. The size of the farm and the amount of equipment are closely related to the possible profits. A farm of forty or eighty acres devoted to diversified crops and live stock, cannot be so economically equipped per acre as a larger farm. The investment per acre in machinery will be higher as the cost will be spread over fewer acres than the machinery has capacity to handle. Investment in other equipment will also be correspondingly high. Often the labour on such a farm is not fully employed and loss results from inactivity of labour and equipment. A medium to large sized farm, when well organized, fully equipped, and with sufficient capital reserved to operate it well, will pay a much better labour income than a small farm.

Proportion of Investment in Machinery

Investments in machinery are worthy of quite as much consideration as investments in land. Machinery is looked upon as one of the means of reducing the cost of production. The wise use of machinery saves time and labour and enables the farmer to handle large acreages. In this light, the use of ample machinery is wise. The fact remains, however, that investments in machinery are often poorly made and that many farmers are embarrassed by debts for machinery. Frequently, farmers purchase machinery because it is fashionable or because a neighbor has it, rather than because carefully made calculations show that a certain machine can be used profitably.

A safe rule is to buy no machine until carefully made calculations show that the cost of production of a certain crop or product will be reduced sufficiently by the purchase to cover the cost of the machine. That machinery investments can be studied from a business standpoint is quite plain. The following example, showing the relative cost of cutting corn with a machine versus cutting by hand, will illustrate:

The original cost of a corn binder is $125. The annual depreciation, as shown by statistical records covering ten years' work with farmers in Minnesota, is $12.50. The interest on the investment at the average value of the binder throughout its life will be $4.12. Repairs, shelter, and insurance will cost $2 annually. The total annual cost for the use of the binder, therefore, will be $18.62. If only twenty acres of corn are grown each year, the annual cost an acre for the corn binder will be $.93. The cost of cutting corn would be as follows: It will be noted that a twenty-acre corn field can be more economically harvested by hand where labour is available. On a ten-acre field, the difference in expense would be still greater in favour of the hand harvesting, as the cost of machinery per acre would be doubled. The scarcity of labour, however, and the necessity of harvesting corn quickly to save it from frost, would often warrant the expenditure for the machine, even though it does slightly raise the cost of harvesting per acre. The greater the acreage, the more useful the machine becomes in harvesting, and the less the expense per acre for machinery use.

Calculations similar to this should be made before purchasing a machine for any purpose. If it can be shown that the cost of performing the labour may be reduced by the machine, and that the labour can be performed in a more satisfactory and efficient manner with the use of it, then the purchase may be warranted. In many cases, however, calculation will show that the purchase is not warranted and that it would be better to hire labour and rent the machine, or to buy in partnership with some one else.

Cost of Motive Power

Another factor that should receive consideration is the cost of motive power in use on the farm. Horses usually furnish the farm motive power, though tractors are used to advantage in some cases. Auto trucks and automobiles can often be used to advantage in marketing dairy, fruit and garden products and the cost of using them, when it can be ascertained, should be compared with the cost of horse power. It costs from $40 to $100 per year to keep a work horse, depending on the locality and on the price of feed stuffs; also on the work that the horse does.

The average cost a year of keeping a horse in Minnesota for the years 1904 to 1907, varied from $75.07 at Halstad, to $90.40 at Northfield.* Often a large number of horses are kept because the farm has been devoted to grain raising and the horses are needed at seeding and harvest times. They run in the pasture during the summer and are idle during the winter months. They must be fed and cared for during this time. The money invested in them would be drawing interest if invested somewhere else. Farmers should reduce the horses kept to the number actually required to do the work, unless the surplus are colts growing in value as one of the market products of the farm. The work of a farm can often be lessened by adopting a good crop rotation and using such crops as do not demand large amounts of horse labour at the same time. In this way a farm of 160 to 240 acres can often be worked with four to six horses, whereas eight to ten are frequently kept. The support of two or three extra horses per year would amount to $200 to $250 and is an item well worth saving.

Investment in Buildings

Investment in buildings, fences, and other items of equipment should be considered in the same business-like way. A barn costing $4,000 and providing shelter for forty head of cattle would carry with it an annual cost of $440. This annual cost is made up from the interest on the investment, insurance, depreciation, paint, and repairs. It will be about 11 per cent on the total investment.

If the same forty cattle could be housed in a barn costing $2000 the total annual cost would be only $220, charging the same interest and expense rates as in the first instance, and assuming that the rate of depreciation would be the same. While the $4000 barn would undoubtedly be a better barn, it would not add to the production of the cows housed, unless it was much more comfortable. It would not add to the net profit from the investment unless the labour of doing the chores and caring for the cows would be considerably reduced by greater convenience.

The cost of horse barns, swine barns, and other buildings can be similarly calculated. One should not erect a building unless it is going to add to the efficiency of the live stock, shelter hay or machinery, or lessen the labour of doing the chores. It is wiser to invest money in drainage or in better tillage of the soil, than to invest it in buildings that shelter unproductive stock, or that add nothing to the earning power of the farm.

Cost of Labour

The employment, organization and direction of labour demands considerable study. The value of a good farm manager lies quite as much in his abilitiy so to select and direct labour as to yield a profit, as it does in his ability to drive a good bargain or sell his crops well.

The only reason for employing labour is to increase the product and consequent profit. If a farmer can, by employing a man eight months in the year at $40 per month, increase the product of his farm by $500, he will be warranted in employing the labour. If, however, the $320 invested in labour should yield an increase of only $200 in the products of the farm, employment would be at a loss.

The Factors of Production

Three primary factors are necessary in agricultural production. These are capital, land, and labour. The adjustment of these three factors is an important part of the business of the farm owner or manager, and determines largely the profits that may be made from the individual farm.

Capital, as commonly understood, includes the money value represented in the investment of the farm property, no matter what the form may be. Implements, live stock, teams, buildings, and other articles of equipment, are each a part of the capital of the farm. Cash for operating is also included. Land represents the larger part of capital on most farms, and demands special consideration because the amount of land available for agricultural purposes is limited. Farmers have for this reason, regarded it wise to secure large quantities in localities where it was cheap, anticipating a rise in value. Location and demand for land in particular sections has led to much speculation, and land values fluctuate frequently. The proportionate investment in each of the three forms,—circulating capital, land, and labour—bears a vital relation to the profits possible from the farm, and must be given the most careful consideration by the person who is buying and equipping a farm.

2

Conceptualization of Technological Change in Agriculture

The environmental impact of agriculture varies based on the wide variety of agricultural practices employed around the world. Climate change and agriculture are interrelated processes, both of which take place on a global scale. Global warming is projected to have significant impacts on conditions affecting agriculture, including temperature, precipitation and glacial run-off. These conditions determine the carrying capacity of the biosphere to produce enough food for the human population and domesticated animals. Rising carbon dioxide levels would also have effects, both detrimental and beneficial, on crop yields. The overall effect of climate change on agriculture will depend on the balance of these effects. Assessment of the effects of global climate changes on agriculture might help to properly anticipate and adapt farming to maximize agricultural production.

At the same time, agriculture has been shown to produce significant effects on climate change, primarily through the production and release of greenhouse gases such as carbon dioxide, methane, and nitrous oxide, but also by altering the Earth's land cover, which can change its ability to absorb or reflect heat and light, thus contributing to radiative forcing. Land use change such as deforestation and desertification, together with use of fossil fuels, are the major anthropogenic sources of carbon dioxide; agriculture itself is the major contributor to increasing methane and nitrous oxide concentrations in earth's atmosphere.

Technological change is among a small handful of topics in what is now referred to as the sociology of agriculture — along with part-time farming and analysis of the farm family — that have received attention during all three major phases in rural sociology.

From the 1930s through the 1950s there were several major studies by sociologists of the impacts of technological change, most of which centered on the rapid mechanization of southern agriculture that began during the Great Depression.

The concern of many of the early rural sociologists with the socioeconomic consequences of technological change, however, largely lapsed during the succeeding phase of rural sociology in which the diffusion-adoption perspective was dominant. In the era of the "new sociology of agriculture," we again see a rekindling of concern with the social consequences of technological change. Moreover, interest in the consequences of technological change has become expressed in more social structural terms and, as demonstrated later, has become more closely integrated with analyses of the genesis of new agricultural technologies.

Several other important observations can be made on the rural sociological literature on technological change in agriculture over the past decade. First, mechanization has received the lion's share of attention; technological change in the forms of improved varieties of crops and breeds of livestock or of agricultural chemicals has been the focus of only a few major ex post studies. Second, as Berardi has observed, there are many problems in decomposing the effects of technological change on farm structure from the effects of other variables such as public policy or changing wage rates; therefore, rural sociologists have often relied heavily on the primary research of others (*e.g.*, agricultural economists and geographers with complementary expertise in assessing techno logical impacts) when developing comprehensive assessments of the socioeconomic impacts of technological change.

Third, the conceptualization of technological change in agriculture by rural sociologists has been heavily influenced by the work of the agricultural economist, Willard Cochrane, who developed a theory of the "treadmill of technology" in the late 1950s. Briefly, Cochrane's theory of technological change builds on a key observation from rural sociological research on the diffusion-adoption of innovations-that large farmers who are entrepreneurial and nonrisk-averse tend to be the first to adopt new agricultural technologies.

By adopting early, these farmers enjoy "innovators' rents" because of their ability to reduce their average per unit production costs. Cochrane posits that after these innovators and early adopters have employed the new agricultural technology, aggregate output of the particular commodity involved will begin to increase and prices will decline disproportionately (because agricultural commodities tend to have a low price elasticity of demand).

Declining prices then begin to force nonadopters to utilize the technology. These subsequent adopters gain very little from the technology; they adopt it merely to be able to stay in business. Slow adopters or nonadopters will tend to be forced out of agriculture because they can no longer compete due to their high average costs of production. Cochrane considers the technological change process to be a "treadmill" in that the majority of farmers, who receive little or no benefit from the technology, are nonetheless forced to continue to adopt new agricultural technologies in order to stay in business.

Cochrane has also observed that the benefits of technological change tend over the long term to be realized in the form of rising land values, particularly if there are commodity programs in effect to cushion a commodity sector from declining crop prices caused by increases in output. The benefits of technological change (innovators' rents from early adoption) tend to be reinvested in farm land, causing increases in land prices and benefiting large landowners over smaller property owners and tenants. These effects, in Cochrane's view, are heightened when commodity programs support crop prices, thereby increasing the level of innovators' rents and intensifying the extent to which these benefits are capitalized in asset values.

Cochrane's theory of the treadmill of technology has been widely accepted in recent rural sociological work as an orienting perspective because it has effectively linked knowledge from diverse origins (rural sociological research on diffusion-adoption, agricultural economics research on technological change, and so on). Cochrane's theory, however, has its limits — in particular, its inapplicability to industrial agriculture, as noted by LeVeen.

Moreover, the notion of the treadmill of technology gives little attention to the origins of technology, that is, to why technologies that have displaced labor or benefited larger operators over smaller ones have been developed, a topic that, has recently come to be a major research issue for those working in the sociology of agricultural science. Nonetheless, with the exception of sociologists of agriculture who have focused on industrial agriculture, rural sociologists have yet to develop a distinctly sociological theory of technological change and its socioeconomic consequences.

While rural sociologists have not yet developed a comprehensive and distinctly sociological theory of technological change and its consequences applicable to both family farming and industrial agriculture, the new sociology of agriculture as applied to technological change has made some significant advances. Several of the most important of these advances are summarized below.

Among the most significant research papers on the topic of technological change in agriculture were two published in the late 1970s that dealt with limitations of the classical diffusion-adoption model. The first, by Pampel and van Es, was an otherwise traditional adoption study that included, in addition to "commercial-profitable" innovations that were the major focus of previous diffusion-adoption studies, a set of "environmental innovations" such as contour farming, terracing, sod waterways, and use of rotation cover and reduced tillage. Pampel and van Es discovered that the correlates of adoption of environmental innovations were far different than those of commercial innovations. For example, while large-scale farmers were observed to be the earliest adopters of commercial technologies, there was no significant relationship between farm size and the adoption of environmental innovations.

They concluded their study by cautioning rural sociologists that the diffusion-adoption tradition, because it had ignored dimensions of innovative behaviour such as the adoption of environmental-conservation technologies, "may have provided within rural sociology a field of knowledge with a narrow empirical foundation on which to base its generalizations". They also emphasized that given the tendency of most environmental innovations to be unprofitable, it was not clear that voluntary strategies to achieve soil conservation and related environmental goals would be successful.

Writing two years after the Pampel and van Es study, Goss focused more systematically on the limitations of diffusion-adoption research. The principal failings of diffusion-adoption research, in Goss' view, were (1) that it was uncritical, if not promotional, towards patterns of technological change that may be detrimental to the interests of society and/or farmers, and (2) that the focusing of research effort on the adoption component of the technology change process had caused rural sociologists to ignore the consequences of technological change for various groups in society.

Goss summarized the early 1970s literature critical of the Green Revolution and pointed out that the Green Revolution had been initiated with the same promotional orientation towards technology that had predominated during the heyday of the diffusion adoption tradition in U.S. rural sociology. He noted that many social scientists who had once worked within the diffusion-adoption tradition in international agricultural activities had begun to recognize the limits of the adoption paradigm.

He argued, however, that this point had not been well recognized in domestic rural sociology and that rural sociology must begin to develop a more comprehensive programme of research on the consequences of technological change in agriculture.

In a manner similar to Goss' exegesis of the Green Revolution literature in terms of the impacts of technological change on social groups and classes in agriculture, Stockdale, in an influential article, raised comparable issues in terms of the ecological aspects of modern agricultural technology. Noting that the contribution of technological change to American agricultural productivity had often been held up as a model for other nations to emulate, Stockdale argued that social scientists had tended to ignore the adverse ecological impacts of energy-intensive technology in the United States and elsewhere. Like Goss, Stockdale called for an expanded programme of research on technological change in agriculture that would emphasize the social and ecological consequences of new agricultural technologies.

Among the most innovative thrusts in sociological research on technological change in agriculture have been the series of books, monographs, and articles published by Friedland and associates. Friedland and his colleagues have focused on technological change in industrial agriculture in California and have identified a number of components of the technological change process that have tended to be ignored in the earlier rural sociological literature. As noted earlier, Friedland has focused on several commodity systems (chiefly tomatoes, lettuce, and grapes) and has found that the social organization of each commodity system — and hence the structural context of technological change — is distinctive.

Nonetheless, Friedland and his associates have demonstrated that the technological change process in industrial agriculture generally is heavily influenced by considerations of the cost of, access to, and control over labor and by the relationships of grower organizations with public and private components of the agricultural research system. They have also demonstrated that the nature of the technological change process and its consequences depend upon the structure of the commodity system. For example, in the aftermath of the termination of the bracero programme in 1964, a mechanical tomato harvester technology was widely adopted in California, displacing three-quarters of California tomato producers and the bulk of their agricultural wage labor forces. In lettuce, however, grower-shippers were able to utilize the threat of lettuce harvest mechanization to discipline the labor force and to avoid the costly process of harvest mechanization.

A further component of the recent rural sociological literature on technological change, one to which Friedland and Friedland and associates have made major contributions, is social impact assessment of new agricultural technologies. Social impact assessments of new agricultural technologies are a distinctive type of research in that they often tend to rely primarily on secondary data or on the results from the published literature rather than on primary research and data collection; this is particularly the case if the assessment must be prepared over a short period of time.

Other distinctive aspects of social impact assessment of new agricultural technologies are, first, that the assessment is ex ante (rather than ex post, as in most previous rural sociological research on technological change) and, second, that the orientation towards these emerging technologies tends to be questioning and skeptical, rather than accepting and promotional. Rural sociologists who have conducted social impact assessments of new technologies have approached their subject matter from the vantage point that most new technologies tend to involve an unequal distribution of costs and benefits and that advance projections of these patterns of the distribution of costs and benefits can enable social scientists and affected groups to redress these distributional impacts more effectively.

An example of the type of knowledge base that has been developed, in part, as a building block for social-impact assessment research is Berardi's comprehensive review of the impacts of agricultural mechanization. Berardi reviewed the re sults of over 180 studies of agricultural mechanization in the United States and elsewhere and summarized the major empirical regularities in studies of a variety of types of mechanization in a diversity of commodity systems. She concluded by pointing out that the rich literature on agricultural mechanization — most of it prepared by nonsociologists — is sufficient so that sociologists may contemplate conducting rigorous ex ante studies of the impacts of new mechanical technologies in agriculture.

Unfortunately, as noted earlier, sociological knowledge on the socioeconomic impacts of mechanization is far greater than that on "biochemical" technologies (plant varieties, fertilizers, insecticides, herbicides, and so on). There have only recently been sociological studies that have delivered on Stockdale's and Berardi's admonitions in the area of biochemical technologies in plant and animal agriculture. It is likely, however, that the great attention that has already been paid by rural sociologists to emerging biotechnologies in agriculture will result in a sustained research programme into the socioeconomic consequences of these new technologies.

It should be noted in closing, however, that while there is great merit in inventorying the results of existing studies on various facets of technological change — as, for example, Berardi has done — sociologists should exercise caution, as technological change should not be assumed to have invariant consequences across time and space. In particular, as noted above, there have been some major recent changes in the structure of U.S. agriculture that one might expect will have significant effects on the technological change process and its consequences.

For example, instead of technological change tending to result in the decline of small farms, as it did during the bulk of the post-World War II period, it may be quite likely that new agricultural technologies such as biotechnology-derived plant varieties, growth regulators, and bovine growth hormone may tend to result in the disproportionate demise of medium-sized, full-time family farmers. Thus future rural sociological research on technological change in agriculture must be cognizant of ongoing changes in the structure of agriculture and how these changes affect the types of agricultural technologies that are employed, the processes of technological change, and their consequences.

RURAL SCOEITY OF AGRICULTURAL SCIENCE

Until the early 1980s, only small communities of rural sociologists were interested in the sociology of agricultural science. This was to change dramatically, however, in the aftermath of a workshop on biotechnology in the summer of 1982 at the University of Kentucky organized by B. Koppel, Busch, and Lacy. Since that conference there has been growing attention to a variety of facets of biotechnology as key focal points of research in the sociology of agricultural science.

This research on biotechnology now represents a large literature, and the approaches developed therein have begun to substantially influence how rural sociologists are exploring other types of agricultural technology such as information technologies. This new line of research has been salutary in yet another respect; sociologists interested in biotechnology have begun to make major progress in uniting the sociology of agricultural science with sociological research on technological change and social impact assessment of new agricultural technologies.

The wide-ranging and potentially integrative character of research on biotechnology, agricultural research, and agricultural change is amply illustrated by rural sociologists having given major attention to no less than 10 major themes under this rubric. With regard to the first, the politics of the evolution of the public agricultural research system, Buttel and Kenney and Kloppenburg found that the basic science discoveries (especially recombinant DNA) relating to biotechnology, the promise of biotechnology to agribusiness firms for developing new input commodities, and federal government concern with the U.S. competitive position in the "high technologies" have rekindled scrutiny of the land-grant and ARS systems for placing too little emphasis on basic biological research. The result was a changing political climate of public agricultural research that has led to rapid expansion of biotechnology programs in the public agricultural research institutions and to a deemphasis on traditional, applied research.

These changes in the research emphases of public agricultural research institutions have stimulated a second area of research in the sociology of biotechnology: industry-university relationships and the changing division of labor between public and private research. Buttel et al., for example, have found that industrial funding of agricultural research at the land-grant universities has increased significantly and this has generated new concerns that agribusiness firms are acquiring excessive influence on research priorities in public research institutions.

Lacy and Busch have identified a large number of likely impacts of closer industry-university relationships in agricultural biotechnology: further decline of public plant breeding (especially varietal releases), greater concentration of scientific talent at a small number of land-grant universities, decreased autonomy of public researchers in establishing their own research agendas, increased secrecy, and greater restrictions on the free flow of scientific information.

A third area of research relating to biotechnology has been on agribusiness reorganization. Kenney, Kloppenburg, and Buttel et al. have noted that the rise of biotechnology research has reinforced or led directly to major structural reorganization in the agribusiness industries. With the passage of the Plant Variety Protection Act of 1970, the ability of private firms to acquire patent-like protection for new crop varieties increased the perceived profit potential of seed companies, leading to the incorporation of most major independent seed companies into multinational corporations as wholly owned subsidiaries or joint-venture partners. Biotechnology reinforced this trend, since seeds are projected to become the major commercial vehicle for plant biotechnology; moreover, the basic sciences underlying biotechnology (molecular and cell biology) are quite similar to those that will ultimately be applied in other areas of agricultural inputs such as agricultural chemicals, innoculants, and animal pharmaceuticals. Thus biotechnology has increased the benefits of uniting chemical and plant breeding product lines within integrated multinational firms that manufacture several agricultural input products. Kenney has given particular attention to the role of small venture-capital or start-up companies in biotechnology and has analyzed the processes by which they have developed linkages with large companies.

Goodman et al. have stressed that biotechnology will have two distinct applications in agriculture — "appropriationism" and "substitutionism" — with different consequences for structural reorganization of the food and agricultural industries. By "appropriationism" Goodman et al. mean the use of biotechnology to develop new inputs, such as new crop varieties and animal breeds, for agricultural production. Appropriationist applications of agricultural biotechnology will tend to result in the integration of agricultural input manufacturing with comparable product lines in other life science industries such as chemicals and pharmaceuticals.

"Substitutionism," on the other hand, refers to the use of biotechnology to substitute for existing products. Substitutionism will therefore have the effect of severing existing production "chains," in which specific raw materials are employed to produce particular final products (*e.g.*, substituting single-cell proteins for vegetable proteins as animal feed supplements). The implications of "substitutionist" applications of biotechnology will also include shifting food and fiber production from the farm to the factory (*e.g.*, substituting industrial tissue-cultured thaumatin for cane sugar). Substitutionist applications will tend to lead to integration of the food and fiber processing and chemical industries.

Fourth, several researchers have examined how biotechnology has affected and will continue to affect plant breeding, including but not limited to public-sector plant breeding. Kloppenburg has conducted an extensive historical analysis of plant breeding and the seed industry in the United States. He has argued that the trends under way as a direct and indirect result of biotechnology promise to increase the penetration of agribusiness interests into land-grant university plant breeding programs and to complete the shift, under way since the late 1930s, to curtail the role of land-grant plant breeders in developing public-domain finished plant varieties. Hansen et al. have reached similar conclusions on the basis of interviews with plant breeders and biotechnology researchers across the public and private sectors. Hansen et al. have noted as well that the increased importance of biotechnology in public-sector plant breeding programs, combined with new avenues for agribusiness influence, will tend to result in the demise of public plant breeding efforts for minor crops.

A fifth area of inquiry in biotechnology, agriculture, and agricultural research has been on the role of intellectual property restrictions and other proprietary considerations in affecting research priorities and the nature of the new technologies that are developed. Kloppenburg, for example, has called attention to private-sector plant breeding firms' longstanding concern about the perceived inadequacy of intellectual property institutions to protect their investments in plant breeding. In the absence of such restrictions for sexually reproduced crops prior to 1970, most seed companies focused primarily on hybrid seeds (largely hybrid corn), which, because of their reproductive instability, afforded a "natural," biological protection against farmers saving their own seeds and circumventing annual purchases of seeds from private companies. Kloppenburg has noted, however, that the elaboration of new legal means of privatizing plant research has increased corporate interest in plant biotechnology. Kloppenburg, Kenney, and Hansen et al. have suggested that the role of intellectual property restrictions, combined with the fact that most agricultural input firms have both seed and chemical product lines, may lead to the development of "seed-chemical packages" that tie the farmer more directly to purchased, capital intensive inputs.

Sixth, Kloppenburg and Buttel have begun to explore the relationships between the public and private sectors in terms of recently developed "theories of the state". They have begun this task by conceptualizing public agricultural research and government commodity programs as reflecting contradictions of state intervention.

They note public agricultural research with state intervention has tended historically to be expanded during times of socioeconomic crisis. Agricultural research, however, can be seen to set the stage for future crises in the agricultural economy by creating long-term increases in agricultural productivity. They discuss the emergence and resolution of these contradictions in the context of the Great Depression and the current era of biotechnology and farm crisis.

A seventh area of research pursued within the rubric of biotechnology has been the implications of this new trajectory of research for genetic resources.

Kloppenburg and Hansen et al. have argued that the increased role of intellectual property restrictions combined with the new opportunities for private firms in developing new purchased input commodities will tend to lead to a decline in germplasm flows between the public and private sectors. In the absence of greatly expanded public and quasi-public appropriations for germplasm preservation, these trends are seen to threaten the genetic resource base and perhaps to constrain plant improvement in the future. Kloppenburg and Kleinman have also examined the forces, including the rise of biotechnology, that have led to North-South conflict over control over crop germplasm.

Eighth, as noted earlier, considerable attention has been focused on how new biotechnology-related inputs such as bovine growth hormone, herbicide-tolerant crop varieties, nitrogen-fixing cereal grains, and so on might affect the structure of agriculture in the future. Kenney et al. have emphasized that agricultural biotechnologies are potentially quite diverse and that how these new technologies will affect farm structures depends heavily on the research priorities that are established.

Molnar and Kinnucan, basing their work on existing theories of and evidence on technological change in agriculture, argue that larger farmers will likely be the first to adopt — and thus be the major beneficiaries of — the new biotechnologies. Kloppenburg has developed a preliminary ex ante assessment of how new biotechnologies applied to corn might affect U.S. farm structure over the coming decades. Buttel has focused on the first major biotechnology to be applied in agriculture: bovine growth hormone, which can be produced in large quantities by genetically modified bacteria and which increases milk production per cow by up to 25 percent. He has argued that this biotechnology will likely reinforce longstanding structural trends towards a dualistic farm structure. The Molnar and Kinnucan anthology contains several important papers on ex ante assessment of new agricultural biotechnology inputs.

A ninth area of biotechnology-related research has been in providing preliminary assessments of how biotechnology might affect the nonfarm economy of nonmetropolitan areas. Finally, Buttel et al., Kenney and Buttel, and Koppel have done research into the implications of biotechnology for international agricultural research and agrarian transformation in Third World countries.

The sociology of agricultural research, especially its recent focus on biotechnology, has been conducted within the "new sociology of agriculture" tradition to a greater extent than any other area of inquiry, save perhaps for the state and agriculture. In particular, the sociology of agricultural research, although involving some quantitative analysis, has been at the heart of trends in the new sociology of agriculture that have created legitimacy for exploring issues that are largely accessible only through historical-qualitative research methods. Especially in biotechnology, the impacts of which on farm structure will generally not be manifest for a decade or more, quantitative research based on primary data is virtually impossible, and the quantitative research that has been done often suffers from spurious precision.

Given the importance of understanding the likely impacts of these new technologies, the work that is being done is of high priority. Some rural sociologists, however, will feel less than comfortable with a research tradition in which quantitative methodology and the putative precision it affords are largely absent. Nonetheless, the sociology of agricultural research and biotechnology has been among the most innovative and stimulating areas of recent research in rural sociology and will no doubt have a major impact on rural sociological inquiry in the decades to come.

Agriculture and the Rural Community

Perhaps in no other area of the "new sociology of agriculture" is there so much apparent continuity with the pre-Korean War tradition of rural sociological research as in research on farm structure and rural communities. The early rural sociologists primarily conceived of agriculture as one of the major institutional ensembles affecting rural community life and, especially during the 1930s, devoted much of their research effort relating to agriculture to understanding the implications of agricultural structural changes for rural community life and viability.

The conceptualization of agriculture and rural community interactions and the methodologies for exploring these interactions in pre-Korean War and post-1970 rural sociological research, however, exhibit dramatic differences. Whereas most of the early rural sociologists tended to be less interested in agriculture per se and were principally concerned with understanding rural community institutions, contemporary researchers have tended to be more interested in structural change in agriculture and have deemphasized understanding the totality of the fabric of rural social life.

The research tools and data sources of these two generations of rural sociological researchers are also dramatically different. The early rural sociologists relied on rural social surveys, while the bulk of contemporary research is based on county-level census data. A further difference between the early and contemporary rural sociologists of agriculture in their work relating to rural communities concerns the pivot of the latter studies around Goldschmidt's "As You Sow". This is somewhat ironic in that Goldschmidt's study was anthropological and qualitative, whereas the research conducted since 1970 that has drawn on Goldschmidt's work has largely been quantitative and has involved far less "anthropological" detail on rural communities than did Goldschmidt's.

Rodefeld is generally credited with the "rediscovery" and popularization of Goldschmidt's pioneering work among rural sociologists, though Heffernan (1972) made important contributions to elevating "Goldschmidt-style" research to a prominent position in rural sociology. Nonetheless, after Rodefeld's refocusing of rural sociological attention on the Goldschmidt thesis, Goldschmidt's book has been virtually an obligatory citation in research articles on this topic. Despite some trenchant criticisms of Goldschmidt's perspective and research methodology, Goldschmidt's general hypotheses remain accepted theoretically and thus continue to shape, if not dominate, research on farm structure and rural communities within the "new sociology of agriculture."

Briefly, Goldschmidt's thesis was that there are differences in the quality of rural community life depending upon the social organization of agriculture and its attendant occupational structure. Goldschmidt observed that the quality of rural life in two carefully matched California communities in the San Joaquin Valley differed dramatically and that these differences were largely accounted for by the organization of agriculture.

He argued that Dinuba, a community surrounded by "family" farms, had a greater number of businesses, a larger retail sales volume, a higher per capita income, and a greater diversity of social, educational, recreational, and cultural institutions than did Arvin, a community surrounded by large, industrial, corporate farms. These differences were principally accounted for by an intervening variable — community occupational composition — that was seen to be strongly influenced by the social organization of agriculture and which directly contributed to the dramatic variations in indicators of the quality of life. Goldschmidt reported that the high ratio of laborers to other occupational groups in Arvin accounted for its low level of community viability relative to Dinuba. From these observations Goldschmidt inferred that the rise of large-scale corporate agriculture then underway in California and elsewhere would have adverse implications for rural communities.

Goldschmidt's study received essentially no further attention until it was referred to in a 1968 hearing on corporation farming before the Senate Select Committee on Small Business. LaRose and Sonka and Heady soon reported further data consistent with the Goldschmidt study. In addition to Rodefeld's and Heffernan's efforts to resurrect Goldschmidt's work in rural sociological circles, there was a major research effort by the Small Farm Viability Project in California, drawing largely on Isao Fujimoto's unpublished research, which presented a larger restudy of the Goldschmidt thesis. The results largely supported Goldschmidt's findings of over three decades earlier. Flora and Conboy also reported results consistent with the thesis.

Since that time there have been numerous studies in the new sociology of agriculture tradition that have explored the Goldschmidt thesis in a number of contexts. Several observations can be made on the conduct of recent research in the Goldschmidt tradition, what we might call "neoGoldschmidt" research. First, it must be stressed that contemporary rural sociological researchers have taken considerable liberties with Goldschmidt's formulations. As noted earlier, Goldschmidt's method was a comparative anthropological case study of two communities and eschewed quantitative analysis of the sort that has predominated in the subsequent rural sociological literature.

Also, Goldschmidt's analysis focused on several aspects of the property relations and control over labor, such as absentee ownership and use of ethnic minority laborers, rather than farm size, which has been the most important farm structural indicator in the neo-Goldschmidt tradition. In that regard it is important to stress that the "family" farms of Dinuba were by no means "small" farms; in fact, the family farms of Dinuba were generally larger in gross sales terms than most large farms in the Midwest today.

Second, neo- Goldschmidt research has largely focused on a single direction of causality — the impact of agricultural structure on rural communities — and has largely ignored other types of interrelations between agriculture and rural communities. Third, most studies have been based on Census of Agriculture data and have employed countylevel data as a proxy for community characteristics have utilized survey data to infer community quality of life consequences of changing agricultural structures. Fourth, despite obvious major regional variations in farm structures and in rural communities and their economies, there has, until very recently, been a virtual lack of a comparative perspective on farm and community structures.

Goldschmidt and Harris and Gilbert are partial exceptions to this generalization, since they have done research using states as the unit of analysis, though this procedure compounds the aggregation bias problems. Very recently, nonetheless, there has appeared a series of regionally based studies of the effects of farm structure on rural community viability and quality of life commissioned by the Office of Technology Assessment. These studies are Buttel et al., Skees and Swanson, Flora and Flora, MacCannell, and van Es et al. as published in abridged form in Swanson. These five studies were very briefly summarized in the main report. Together they indicate strongly that there are pronounced regional variations in how changing farm structures affect rural communities.

The impacts of farm structural changes on rural communities are generally most pronounced in the Great Plains and industrial agriculture states, less so in the South and Midwest, and very little in the Northeast. There is evidence that the degree to which farm structure affects rural communities is directly associated with the dependence of the rural community on agriculture.

The general model pursued in these neo- Goldschmidt studies has been to estimate coefficients for a causal chain in which farm structural change indicators are seen to affect the composition of the farm population (*e.g.*, as measured by indicators of the size of the farm population, the proportion of full-time hired workers) and the size and composition of the rural population. These clusters of variables are then seen to affect indicators of the quality of life in rural-agricultural communities (*e.g.*, proportion of families in poverty, median family income, per capital retail sales, unemployment rate).

This general model is sometimes supplemented by technological factors as antecedents of farm structure, since technological change is considered to be one of the major factors influencing farm structural change. More recently, studies in this genre have gone beyond the cross-sectional census data designs that predominated from the mid-1970s to the early 1980s and have included measures of change, where appropriate, in variables in the model. One limit to studies that use census and other data based on single-year assessments of farm income is that net farm incomes are highly variable over the short-run; for example, as many as 20 percent of farms with annual sales of over $250,000 had poverty-level incomes in 1986.

Heffernan has effectively summarized the general configuration of results in the neo-Goldschmidt literature by noting that "all relevant research to date suggests that a corporate type of agriculture results in a reduction in the quality of community life for at least some people, especially the hired workers in rural communities." As such, the results continue to lend support to the Goldschmidt thesis. But there have recently been several expressions of dissatisfaction with Goldschmidt's perspective and with the generalizations that have been made from research in this tradition.

A major theoretical concern is that Goldschmidt and many researchers working in this tradition have taken large-scale and corporate agriculture to be basically synonymous when, in fact, not all large-scale farms are corporate or industrial-type farms. Much more important than industrial-type farms in farm numbers and shares of sales, assets, and profits are larger-than-family farms. Empirically, as Goldschmidt-style research has been done in the Midwest and other traditional family farming areas, there have been admonitions that large-scale agriculture, especially of a nonindustrial character, does not necessarily lead to adverse impacts on rural communities, though some studies note that increased scale of production could have detrimental impacts on community services.

A further area of dissatisfaction with research on farm structure and rural communities is that the treatment of farm structure in this literature has generally failed to incorporate new concepts relating to agricultural dualism. Typically farm size or scale is operationalized as a single variable representing a central tendency measure of farm size, which was a suitable procedure when there was a singular process of differentiation among farms. But with the emerging dualism in U.S. farm structure since the late 1970s, there has been little change in mean acres per farm since the 1969 census. This measure of central tendency thus ignores the processes of dualism that have contributed to increased concentration with apparent stability in average acres per farm since the late 1960s.

The neo- Goldschmidt conceptualization of an agricultural structure in which particular communities can be characterized along a continuum of farm sizes ranging from small to large has also been questioned by Wimberley who, using multiple indicators from county-level Census of Agriculture data and a longitudinal design, found that there are instead three continua relating to degree of "corporate-commercial" agriculture, "largefarm-area" agriculture, and "small-farm" agriculture. Using Wimberley's measures and including sources of off-farm employment as well, Reif found that large-farm-area agriculture and core-industry structures were related both to family poverty and income inequality in counties. Thus, while studies of the impacts of changing farm structures on rural communities have been a productive area of research in the new sociology agriculture, a number of research frontiers for the future are suggested by recent criticism. There is a need to diversify the methodological approach, consider alternative directions of causality, introduce more sophistication in terms of the measurement of change in farm structure, and expand recent efforts at comparative regional and interdisciplinary research.

The work of Vail represents an encouraging thrust in this direction. Vail has conducted extensive research on the relationships between agriculture and rural communities in Maine in which a variety of factors of demonstrated importance — the reciprocal relations between agriculture and rural communities, the role of an increasingly dualistic agrarian structure, nonfarm aspects of agriculture, and the historically specific nature of farm-community interactions — have been incorporated into a single study.

The notion that American agriculture is no longer (if it ever was) isolated from the remainder of the political economy was driven home clearly by the farm crisis of the 1980s. This crisis was brought by the conjunction of a number of factors in the 1970s, including: accelerated demand for grain on world markets, public policies that encouraged massive capital investments and expansion of production, and a political-economic context that encouraged land speculation. The farm crisis emerged in the 1980s for a number of reasons. Global agricultural export markets shrank, in large part due to the global recession. Farm commodity prices and ultimately net farm incomes (exclusive of federal commodity programme payments and price supports) declined due to the glutting of world grain markets. Reagan administration fiscal and monetary policies led to high real interest rates, which were especially detrimental to agriculture because of its interest rate sensitivity; accordingly, debt-asset ratios among farmers increased rapidly.

The 1980s also witnessed the emergence of new Third World agricultural export competitors (*e.g.*, Argentina, Brazil) because of the global debt crisis and these countries' need to increase exports in order to repay foreign loans. Further, federal fiscal austerity eroded the federal government's ability to maintain farm incomes through commodity programs (even though commodity programme expenditures soared during the decade). The result was a massive decapitalization of American agriculture, especially in the Midwest and Great Plains, as farmland prices in the states most affected by the crisis declined by over 50 percent.

This crisis attracted a good deal of attention by the mass media and by agricultural economists. It was not, however, until the middle of the decade that much was published about the farm crisis by rural sociologists. Most of this research can be divided into four categories: (1) neo-Marxian (and related neo-Weberian) analyses of the continuing penetration of capitalism into and the role of the state in agriculture, (2) analyses of the differences in the characteristics between farm(er)s that were and were not experiencing "financial stress," (3) studies of the effects of farm financial crisis and failure on the farm families whose enterprises were affected, and (4) studies of public reactions to the farm crisis.

The neo-Marxian (and neo-Weberian) approaches have largely focused on understanding continuing cycles of crises in agriculture under capitalism. Although these analyses focus heavily on the political and economic context of agriculture, they are not fundamentally different from what is discussed and therefore are not discussed specifically here. Examples of this genre specifically focused on the crisis of the 1980s include Berlan, Bonanno, Mooney, McIntosh and Zey-Ferrell, Zey-Ferrell and McIntosh, Goodman and Redclift, and Buttel.

We also briefly examine the farm crisis from a methodological angle — namely, the growing tendency in the political economy of agriculture to stress periods of agricultural crisis as key "test cases" for assessing theories of agrarian transition.

Studies in the second category, analyses of the characteristics of farms with various levels of "farm financial stress," deal more with the effects of the political and economic environment of agriculture on farms of different characteristics. These studies have produced inconsistent results due to regional, ethnic, and other variations within the farming population.

For example, farm-size variables were found to be directly correlated with financial difficulty on Iowa farms and unrelated for Texas and North Dakota farms. Yet aggregate national data revealed that larger farms were more profitable, a relationship that obscured the underlying failure of many middle size farms. Farm families with fewer years of farming and younger ages were affected more severely.

However, Murdock et al. found the cohort that had started farming between the end of World War II and the mid1960s also had higher rates of financial difficulties, suggesting a life-cycle effect for those farm families bringing another generation into the farm. The literature does, however, show agreement on the relationship between other variables and financial crisis. For example, higher debt-to-asset ratios brought by low commodity prices in a time of deflation of asset values were associated with financial problems.

The third category, analysis of the effects of the farm financial crisis and failure on the farm families whose enterprises were affected directly, has largely been the work of William and Judith Heffernan, who have conducted in-depth interviews among Missouri farm families who were in severe financial crisis or had been displaced from agriculture. These families experienced a host of difficulties beyond the simple loss of their farms and the disorientation that accompanies the loss of a valued way of life. Among their major results are the following: Many farm families felt socially isolated and were not well-prepared for participation in nonfarm labor markets. Local employment opportunities were often unavailable, and tax laws devastated some families economically after they had attempted to leave farming with some resources with which to start a new life. While the long-term effects of farm failure for these families were not yet known, it was clear that some families were adapting to their new situations successfully while others were at risk of suffering long-term poverty. Graham's study of displaced farm families in New York State bears out these findings. This category of studies has implications for the mutual effects of agriculture and its environment as the displaced farm families affect local labor markets and community social structures.

The final category of research on the 1980s farm crisis, on public reactions to the crisis, is best represented by Lyson's research on public apathy. Drawing on theories of public opinion formation, Lyson demonstrates that several factors led to a surprisingly high degree of apathy by nonfarm people towards the farm crisis, despite the massive media attention the crisis attracted. He notes that there has been a trend towards farmers and consumers having very little direct contact and that the relationship between food prices and farm income has become concealed. Most importantly, Americans have come to view "bad news" from the farm sector — the cost-price squeeze, high debt loads, and the loss of family farms — as being normal. Thus the American public, though "interested and concerned" about the economic woes of farmers, exhibited apathy towards the farm crisis.

A major neglected area in the rural sociology literature on the farm crisis to date is studies of the effects of the farm crisis on rural communities. This topic pertains, in particular, to how conditions in agriculture affect it's environment, especially at the local level. Financial losses to local agribusiness firms when farms fail, losses of farm families in local rural market areas, declining property tax revenues due to deflated farm asset values, and other problems associated with the declining prosperity of agriculture have likely had severe effects on local services and on the local social fabric.

THE SOCIOLOGY OF AGRICULTURAL SCIENCE

We noted several times in the preceding that the era of the new sociology of agriculture was initiated in the aftermath of a distinctly new political milieu of agricultural science and research. Hightower *Hard Tomatoes, Hard Times*, although sensational in tone and misleading on several aspects of the public agricultural research system, attracted a great deal of attention from defenders and critics of the land-grant research system.

Most importantly, Hightower's book (along with contemporary ecological criticism of U.S. agricultural technology) opened up several new lines of social science inquiry into the structure and functioning of the public agricultural research system.

Probably in no other state did public agricultural science become as controversial as in California after the publication of Hightower's book. The California Agricultural Experiment Station has long had close relationships with industrial farmers and other segments of agribusiness and had been held directly responsible in the eyes of many for developing new technologies, such as the tomato harvester, which had resulted in the rapid loss of tomato farmers and the displacement of hired farm workers. It was thus not surprising that the initial sociological inquiries into the conduct of public agricultural science were conducted in California.

The first set of studies by Fujimoto and others at the University of California, Davis, were directly inspired by Hightower and were largely focused on how agricultural scientists made research problem choices. In particular, Fujimoto and colleagues devoted attention to how research problem choices were determined by "internal" factors (such as tenure and promotion considerations) and "external" factors (such as the influence of agribusiness interests and grants). They found on the basis of personal interviews with scientists that external factors were quite important, though internal factors also exerted major influence on scientists' research priorities.

The second set of studies focused on California was initiated by Friedland and associates, who examined the processes by which agricultural commodity groups organized to exert influence on individual scientists, departments, and the University of California system as a whole in order to ensure the development and availability of technologies that would enhance their profitability. Friedland and associates gave particular attention to how the responses of University of California research officials and scientists came to emphasize mechanization and mechanization-related technology that enabled growers to displace farm workers or to enhance the control of growers over their hired labor force.

Based on the results of several years of research during the 1970s, Friedland and Kappel published a short, but still influential, monograph calling attention to biases in the public agricultural research system and outlining mechanisms by which social scientists and groups adversely affected by particular types of agricultural research could achieve changes in the land-grant system.

Shortly after the California studies were initiated, a research group at the University of Kentucky led by William B. Lacy and Lawrence Busch began a programme of inquiry into the sociology of agricultural science that continues to make major contributions to a sociological understanding of the processes by which new agricultural technologies are generated. The first publication of this group, by Busch, was a theoretical critique of the prevailing conception of scientific knowledge by researchers in the diffusion adoption tradition. Subsequent publications by the Lacy-Busch group have focused on a variety of aspects of the structure and functioning of public agricultural research activities.

Two books published by the Lacy-Busch group have had particular impact on rural sociology and the larger social science community exploring agricultural research. The first, *Science and Agricultural Development*, edited by Busch and containing several papers by his research colleagues, was the first book-length treatment of agricultural research in the tradition of the new sociology of agriculture. The chapters in the Busch anthology most germane to the sociology of agricultural science are the following: Randolph and Sachs compared the establishment of medical and agricultural science, emphasizing how a reductionist orientation and technocratic ideology had been developed in each during the process of institutionalization.

Busch and Lacy summarized the existing literature and their preliminary research on sources of influence in research problem choices in the agricultural sciences. Striving to further the social scientific conceptualization of agricultural science and technology as social products, they explored two rival social science views of the social construction and social structure of science. One, the "internalist" view deriving from Merton work, Sociology of Science, essentially holds that the outcomes of the scientific research process are determined by the "norms of science," held by practitioners and reinforced within scientific institutions, and by the very subject matter itself. The other perspective, the "externalist" view, argues that the outcomes of the research process are largely shaped by the institutions, interests, and availability of resources that originate and are determined outside of scientific disciplines and research institutions.

They suggest that both perspectives have a certain validity: "The internal and external positions are best viewed as distinct strata in the life-worlds of practicing scientists. Thus, for each discipline and institution, science is a continuous process of multilevel negotiation regarding the sources of influence on research". A final paper in the Busch collection of particular relevance to the new sociology of agriculture was Busch and Sachs', which employed a world-systems framework derived from Wallerstein to examine the establishment of the agricultural sciences in the First and Third Worlds since the seventeenth century.

Busch and Lacy are known best for their book, *Science, Agriculture, and the Politics of Research* culture, and the Politics of Research, which summarized an ambitious study of research problem choice through the use of historical methods, personal interviews, and sample survey data. Following several introductory chapters that discuss the establishment of public agricultural research, the organization of the public agricultural research system, the social demography of agricultural scientists, and the structure of communication in the agricultural sciences, they explore three major sources of influences on agricultural research: disciplinary, organizational, and extra-organizational. Busch and Lacy found scientific disciplines and disciplinary organizations to be powerful and growing influences over scientists' research problem choices.

Scientific disciplines affect scientists' research choices in a variety of ways, particularly through controlling peer-reviewed journals and establishing the norms adopted by land-grant universities and the Agricultural Research Service (ARS) in tenure and promotion decisions. Busch and Lacy also found dramatic differences among the agricultural science disciplines according to the degree to which they emphasize basic or applied research, the rate at which they accept papers in scientific journals, their norms for scientific productivity, and the degree to which scientists exhibit consensus on what are "hot" topics in the field.

Organizational context — including ARS vs. land-grant university context and the institutional prestige of the land-grant institution — was found to be a significant influence on research problem choices. Busch and Lacy found that ARS scientists were more highly constrained in their problem choices than land-grant agricultural scientists. They also found that institutional prestige was positively related to the degree to which extra-organizational influences and constraints on research choices were manifest. Among the extra-organizational influences, granting agencies and organizations (those hiring scientists on a short-term consultancy basis) were the most direct and pervasive. Whereas there were very large differences among the disciplines in the extent to which their scientists relied heavily on extramural support, Busch and Lacy emphasize that fiscal constraints on public research appropriations are increasing the pressures on scientists to obtain grants and contracts from government agencies, agribusiness firms, foundations, and so on; accordingly, scientists are becoming less autonomous in their research problem choices.

THE CASE OF SOIL EROSION AND CONSERVATION

Environmental problems are by no means new to American agriculture. The Dustbowl episode of the 1930s was but one of several historical instances in which concern has been raised about the environmental performance of U.S. farmers. Nonetheless, current controversies about the management of agricultural resources in the United States date largely from the rise of the environmental movement in the 1960s and have included issues relating to soil erosion, the use of agricultural pesticides, energy and water consumption, and the like. The soil conservation implications of American agriculture have remained among the most longstanding agricultural-environmental issues of concern to rural sociologists and are given principal emphasis.

The majority of U.S. farms are family proprietorships in which there is household or family ownership and management of the land and water resources. Accordingly, the logical point of departure in understanding the soil erosion problem in agriculture has been to conduct micro-level research on the individual, farm firm, and ecological factors associated with soil erosion rates or the utilization of soil conservation technologies. This has been a productive line of research, but it has also become a controversial one, with the major controversies centering around theoretical approach, the relationships between micro and macro levels of analysis, and the public policy inferences that should be drawn from social science research.

Conceptual Issues in Research on Soil Erosion and Conservation

One of the most crucial aspects of soil erosion research is the very conceptualization of why soil erosion is a socioeconomic problem. Traditionally this has not been an issue, since sustained high levels of soil loss were presumed to irreversibly degrade the productivity of agricultural resources.

Accordingly, if the soil erosion problem is conceptualized as medium to long-term land productivity decline detrimental to the interests of farmers, it would follow that conservation should be achieved through voluntary farmer compliance (rather than through mandatory regulation) based on the logic of the long-term interests of farmers.

Historically, federal and state government policy thus has emphasized education of farmers and modest levels of financial inducements (such as Agricultural Stabilization and Conservation Service cost-sharing programs) to lever farmer decision making. But it has been increasingly recognized that the proportion of prime U.S. farmland that stands to be irrevocably degraded by soil erosion is relatively small and that the "off-site" costs of soil erosion — the impacts of soil erosion and run-off on water and land resources of a farmer's parcel — are very significant and may very well be in excess of the on-site costs.

These alternative conceptualizations of the soil erosion problem — productivity loss versus destruction of off-site water and land resources — have very profound implications for appropriate types of social science research and for public policy. To the degree that soil erosion primarily results in land productivity losses, research should focus on individual- and farm-level factors that influence resource management, and public policy should incorporate these research findings in order to develop educational and incentive programs to achieve soil conservation through farmer self-interest. To the degree that soil erosion is primarily a problem because of off-site costs, farmers cannot be expected to conserve because the long-term productivity benefits of conservation are small, and research at a more macro level (*e.g.*, to determine the degree to which farmers' incentive structures and resource management behaviors are congruent with the public's interests in clean, navigable waters) would be most appropriate.

Debate over the appropriate kind of analysis in soil conservation research has been closely related to, and in many respects was stimulated by, debate over the adoption-diffusion approach. Pampel and van Es as noted earlier, reported that the correlates of adoption of conservation technologies tended to be different from those of commercial nonconservation technologies. They also suggested that effective conservation practices will tend not to be profitable for farmers, so that to achieve the public interest in soil conservation voluntary compliance among farmers may be insufficient. The issues raised by Pampel and van Es have continued to pervade the erosion and conservation literature. Many researchers have conducted their research by seeking to revise the microsociological diffusion of innovations approach, while others have argued that the constraints to soil conservation behaviour tend to be more macro in nature and that the diffusion of innovations approach will be inadequate.

Prior to 1970 soil and water conservation technologies and practices were generally understood to be a diverse collection of cropping patterns (crop rotations, contour planting, strip cropping, cover cropping, sod waterways, filter strips) and physical and biomass structures (terraces, sediment retention basins, diversion channels, animal waste structures, and hedge rows). Since that time, however, soil and water conservation has, for the majority of researchers and policymakers, come to be largely conterminous with the adoption and use of no-till and related conservation and reduced tillage equipment, which enables farmers to leave a mulch layer of residue on the soil to reduce runoff and sediment losses.

Two observations can be made with regard to traditional and reduced-tillage equipment approaches to soil and water conservation in agriculture. First, these two approaches, while by no means incompatible, are nonetheless quite different and will tend to apply most appropriately to different types of farm operations. Traditional cropping practices and structures are often most attractive to smaller, more diversified farmers, while reduced tillage tends to be most attractive to larger, more highly mechanized producers of row crops.

Second, reduced tillage, although acknowledged to be effective in reducing erosion in row-crop monocultures, has become somewhat controversial. One controversial aspect of the reduced-tillage approach has been that it requires increased use of herbicides, which may lead to contamination of water and thereby negate some of the runoff reduction benefits of reduced tillage. Many studies have indicated, moreover, that reduced tillage practices have been adopted by farmers largely because of labor savings rather than because of the ability of the technology to reduce soil erosion.

Indeed, the farmers who have been most likely to adopt no-till and other reduced tillage technologies have been found to be large operators with highly specialized operations involving practices such as continuous cropping and row-crop monocultures. Thus farmer-adopters of such conservation technologies ironically represent the major trends in U.S. farm structure that have been demonstrated to lead to environmental degradation in agriculture. However, farmer perceptions of a threat to soil productivity may lead to adoption of soil-conserving innovations, whereas stewardship related motivations may have little effect on motivations to prevent other kinds of consequences such as off-site pollution.

THE ENVIRONMENT AND THE FOOD SYSTEM

Refashioning the genetic structure of crops and animals, important as it is, represents only one face of the technological revolution which has marked the appearance of the modern food system. Just as nature has been redesigned from within, through animal genetics and plant biotechnology, it has been transformed from without, by leading, in turn, to changes in the physical environment, or countryside. These changes in the 'face of the land', themselves stimulated by technological change and accompanying shifts in the labour process have contributed to new social conflicts and political demands around what has come to be seen as the Green agenda. The development of the food system implies a major change in the 'natural' environment, as well as in the genetic composition of the species. The scope of this change can hardly be overestimated.

The environmental effects of modern agriculture point to contradictions within the entire model on which the modern food system is based. Modern agriculture developed out of the partial separation of 'farming' activities from the natural resource base on which these activities were traditionally dependent. This led to technological innovation, and accompanying policy incentives, which have presented problems for the planned evolution of the food system.

In some sectors, such as the production of chemical fertilizers, plant breeding and pesticides, innovation developed 'autonomously', according to a scientific and industrial logic only loosely related to the cycle of agricultural production. The reintroduction of new industrial components into agriculture, and the food system more generally, posed problems in two ways: some processes were taken out of 'farm' production (appropriation) and others were substituted, especially in food processing activities (substitution).

These transformative technological changes were associated with major shifts in the labour process, not only on the farm but in the household and in the food industry. However, an area of equal concern has emerged during the last decade or so, which promises to pose even more serious problems for the evolution of the food system: the environment. Recent decades have witnessed a series of policy interventions designed to reduce, or reverse, the unforeseen or negative effects of the food system on the environment, largely in the industrialized countries. No area of concern demonstrates the difficulty of managing the contradictions of the food system as clearly as the environment.

Conflicts over the rural environment, and over food quality, have come to represent the clearest indication yet of the problems which the modern food system poses for sustainable development, and for the model of economic growth based on intensive accumulation which, as we have seen, spawned the modern food system itself.

This chapter examines the impact of the modern food system on the environment, in both the developed and developing countries, and the social responses which these environmental changes have evoked in recent years. The analysis begins by considering the changes within the food system which have undermined long-term sustainability, changes which were associated with the transference of economic activities between the farm and other components of the wider system. The major agricultural policy instruments which have been brought to bear on the rural environment in the industrialized countries are examined, notably the effects of price-support policies and economic incentives in the introduction of more capital intensive, 'high'-technology agriculture.

The effects on the environment of these policy mechanisms, including the loss of genetic resources, the pollution of water and soils, the threats to sustainable utilization posed by land degradation and the effects of agricultural change on amenity and recreation in rural areas. These environmental effects of the modern food system are considered as first-, second- and third-order problems, each of which requires major adjustments, at the international as well as the national level. Finally, the chapter examines the role of management interventions and environmental movements in seeking to modify or reverse these processes. A conceptual framework is outlined, which explores different forms of 'resistance' to the development of the modern food system, and concentrates attention on the epistemological and political aspects of these responses.

The description of farming in Hooten Pagnell contains most of the elements of what was still, in the 1930s, looked upon as a 'natural' farming system, heavily dependent on the cyclical and seasonal quality of rural life. Animals were left to graze in the fields, or fed on fodder crops produced on the farm, bought inputs such as fertilizers and pesticides played a negligible part in the farm accounts, a sizeable part of the food for home consumption was still provided by the farm, marketing was still in its infancy, and the 'farm family' (largely the women) processed and prepared food in the home. Each of these elements, as we have seen, has been removed (partially or fully) from the farm economy and relocated, usually in an industrial form, elsewhere in the food system.

In contrast to this picture, modern agriculture, as we have seen, is dependent on very high levels of protein and energy transfer, linking the production process on the farm with sectors in which value is added to farm production. These linkages, both upstream and downstream, have broken into the cycle of renewal, failing to return to the resource base what has been removed from it, a principle which marks sustainable agricultural systems. Thus, in modern farming, animal waste disposal has become a 'problem' rather than an essential part of a natural recycling process.

In the same way, advanced systems of crop and animal husbandry, using every protective device from pesticides to antibiotics, are made more necessary by the conditions under which animals are kept, and the genetic make-up of delicately engineered plant varieties. Given the commitment to high yields, consistent quality, and the major investments in research and development which underpin these technologies, once they are established there is apparently no cost-effective alternative to capital-intensive, energy-wasteful production regimes.

Nevertheless, in the process of establishing these regimes the 'knot' which tied agriculture to the environment has been un-done, and the costs of modern agriculture are counted elsewhere. In the United States for example, it has been estimated that the cost of soil erosion off the farm was US$6.1 billion in the mid-1980s. Similarly, in the United Kingdom, £200 million in capital costs and an annual expenditure of £10 million will be required, over a twenty-year period, to bring all drinking water up to the European Community's standard for nitrate concentrations.

In the developing countries the so-called 'externalities' are often higher still. Repetto *et al* (have argued that although gross domestic product (GDP) in Indonesia grew at an annual rate of 7.1 per cent between 1970 and 1984, if this was corrected to take account of the depletion of timber, petroleum and top soil, the rate of economic growth falls to only 4 per cent, and for four of the fourteen years was actually negative. Similarly, by calculating the effects of soil erosion on agricultural growth over ten years, Bishop and Allen have shown that in Mali soil erosion accounts for up to 16 per cent of agricultural GDP. Costs of this sort are not included in the national accounts, and hence calculations about the benefits of agricultural growth rarely reflect the real losses sustained over long periods.

There are several important respects in which attention has shifted to the problems the modern food system presents to the environment. First, the increased dependence on differently located components of the agricultural production and food processing system has served to transfer environmental costs from one part of the system to another. For example, as we have argued elsewhere, the use of soya meal and manioc for intensive cattle production in Europe, has transferred some of the costs of maintaining feedstock systems to other countries, such as Brazil and Thailand, which can ill afford it. Second, increasingly intensive production has led to problems of waste disposal, pollution and disease control, all of them problems which have accompanied intensification.

These problems frequently assume major importance in two directions: for the farm they represent increased costs which need to be incorporated in the farm accounts, and the market viability of the enterprise; and for the local environment they represent major threats to the existing ecological systems, species diversity and habitats.

With increasing demands for financial security, but tighter economic margins, expenditure on what are termed 'intermediate goods' now represents nearly half the value of agricultural output of some industrialized countries such as France.

Table: Intermediate Consumption in French Agriculture (Billion Francs)

	1970	1975	1981
Animal feed	6.8	14.9	33.0
Fertilizers	4.2	9.0	19.1
Fuel and repairs	3.2	6.8	15.4
Crop protection	1.2	2.9	9.6
Building maintenance	0.8	1.4	3.2
Veterinary costs	0.9	1.9	4.5
Other goods	2.5	4.8	9.5
Other services	1.4	2.4	5.0

This table, which expresses the costs of intensification both in current money terms, and as a percentage of total production costs, clearly points to the increasing financial burden of maintaining intensification, which is borne by the farmer in the first instance, but indirectly by the consumer and taxpayer and, at another remove, by agricultural competitors in less protected markets. The environmental costs of this pattern of financial and resource management on the farm are paid in a number of ways, not all of which are easily 'accountable' in monetary terms.

The third aspect of the removal of 'natural' processes from the farm, and the relocation of these processes in other sectors of the food system, has been to transform, in some cases fundamentally, the pattern of land use in rural areas. The modern food system has brought unprecedented interference with traditional land uses and the rural landscape, especially through activities like drainage, land levelling and deforestation.

There has also been a reduction in soil quality, arising from the poor management of soil resources. Current problems, as we shall see, include precipitation of acid rain on land, the discharge of toxic wastes and the uncontrolled spread of sewage on land. As a result of these and other factors, the usable agricultural area within the European Community declined by almost 8 per cent between 1962 and 1984, from almost 100 million hectares to under 92 million hectares.

The combination of several different processes, whose cumulative effect is to undermine the quantity of under-graded land available for both agricultural and non-agricultural uses (including conservation) has led to a rural resource crisis in Europe of serious dimensions. As the Food and Agricultural Organization (FAO) of the United Nations has put it, in rather august language:

A complex web of economic policies provides, at considerable cost to the environment, many of the incentives and institutional mechanisms needed to support the high level and intensity of agricultural production. This situation is compounded by over-production of some commodities in the agricultural sector and rapidly rising economic costs from policies intended to encourage production and raise income. It is to this 'complex web of economic policies' that we need to turn next.

Policy Instruments and Mechanisms

In a document by the European Parliament itself, we are reminded of the lack of attention to long-term sustainability in the development of the European Community's agricultural policy. Emphasizing the five fundamental objectives of the Common Agricultural Policy (CAP) of the European Community (article 39 of the Treaty of Rome)—to increase agricultural productivity, improve farm living standards, stabilize markets and ensure a secure supply of foodstuffs at a reasonable price—the document continues by stating that 'the pursuit of these objectives clearly betokens a productivist and materialist ideology which takes no account of environmental concerns'.

Community preference follows logically from the establishment of a single agricultural market, but it is also clear that establishing this market is expensive in terms of export subsidies 'and that it does not necessarily contribute to the rational location of different types of production'. Similarly, within the European Agricultural Guidance and Guarantee Fund (EAGGF) of the European Community, 'guidance' to farmers (which might assist with the painful process of restructuring for farm households) only amounts to one twentieth of the total expenditure: the bulk of expenditure under this fund takes the form of subsidies to agricultural producers.

The scale of agricultural subsidies, as we shall see, makes any serious attempt at environmental protection in European agriculture extremely difficult, since farmers who are normally subsidized for doing what is harmful to the environment also expect to be subsidized for not doing what is environmentally harmful. The cost of the CAP is, as Simpson remarks 'legendary'.

Only part of the total cost of the CAP is attributable to the purchase, storage and export subsidies for farm production: a 'mere' £18 billion per annum, equivalent to £5.91 a week for a family of four in the Community. In addition, the same family has to pay almost twice this amount weekly towards the cost of fixing EC prices above prevailing world market levels. The total cost is £42 billion per annum. These figures provide an indication of 'a policy which sacrifices long-term interests to immediate concerns, and one scarcely needs reminding that most environmental issues arise from a concern for the long-term future'. It is worth paying attention to the policy framework for agriculture within the European Community, because in intention, although not always in its choice of mechanisms, it mirrors the approach of other industrialized countries, such as the United States and Japan. The policy framework for European agriculture, as we have seen, was established with a view to protecting rural incomes through production subsidies, and ensuring food sufficiency within member countries. One consequence of these policies is that the benefits and costs of agricultural policy are shared very unequally between producers and consumers. For example, if we compare France with the United Kingdom, we can appreciate that because of the larger size of the French agricultural sector, the absolute value of protection is higher than in the United Kingdom.

However, if we take individual farmers the reverse is true: British farmers do rather better under the CAP because they work, on average, larger farms. In France in 1980 the average farm size was 25.4 hectares compared with 68.7 hectares in the United Kingdom. As Cheshire makes clear, 'large farmers, because they produce more, effectively receive more support than poorer, smaller farmers'. The aggregate effect of these distributional factors is highly unequal: three-quarters of EC support to farmers goes to the richest quarter of farmers.

The level of expenditure under the CAP far outstripped most other budgetary items within the control of the Community. Taxpayers and consumers effectively supported 40 per cent of the value of farmers' produce during the years 1983-7, when agricultural spending in real terms increased by 4 per cent a year. In the United Kingdom, in 1984, more was spent on farmers and the food industry (£6.9 billion) than was spent on unemployment benefit and supplementary benefit combined (£5 billion).

Senior also points to the fact that only a dozen people work for the European Commission on consumer affairs, while 800 work in DG.VI, the directorate responsible for agriculture. And within the European Community it is DG.VI that makes agricultural policy, in consultation with the agricultural producers' group COPA, the Council of Agricultural Ministers and the Agricultural Committee of the European Parliament. The effect of close collaboration between farming interests, and the virtual exclusion of the consumers' interest from food policy discussions, is that policy not only favours large farmers over smaller ones, it also favours wealthier families over poorer ones. In 1988 in the United Kingdom the poorest 25 per cent of households spent 29 per cent of their disposable income on food. The richest 25 per cent of households spent 14 per cent on food. The relatively 'rich' West Germans spent 14 per cent on food, while the relatively 'poor' Greeks spent 36 per cent.

Simpson observes that 'if poor people were paying to help other poor people and keep them on the land that may be some consolation. But they are not'. Very little influence is wielded within European Community circles by the consumer interest, and environmental considerations still do not enter into the design of agricultural policy, as one might expect. As we have seen, this lack of political influence is in marked contrast with the position of the agricultural lobby. In explaining the ideological dominance of this lobby, we must take account of the level of farm support in the past. The level of farm support has been fundamental to the shift in agricultural practices, most of which, as we shall see, are prejudicial to the environment.

It is not difficult to demonstrate that the mechanisms, and the level, of farm support in the European Community is highly inegalitarian in its effects. Can it be demonstrated that these policy mechanisms are ultimately responsible for the environmental problems that afflict the countries of Western Europe?

The cumulative effects of high levels of farm support on the rural environment may be complex and subtle, but they are also increasingly well documented. The agricultural technology/policy model encouraged the rapid adoption of new technology in the United States and Western Europe. The high level of farm support, expressed through high farm prices, has encouraged more investment in research and development, and the acquisition of more capital equipment by farmers. Price support has also stimulated intensification through its effect on land values.

Throughout the period between the mid-1950s and the late 1970s agricultural land values increased in the United Kingdom, in fact they tripled in real terms. After 1979, when the rate of increase of CAP price began to fall, and particularly after 1984 when milk quotas were introduced, land prices began to decline, for the first time in decades. The intensification of agriculture was given a considerable stimulus by the capitalization of land as an asset, and the need to maximize returns on this asset.

It is also clear, from evidence such as that of Traill that high support prices for farmers induce more capital investment, and the substitution of capital equipment, where possible, for labour. In West Germany, in 1989, there were 8.5 milking machines for every 100 cows; 2.8 combine harvesters and 12.4 tractors for every 100 hectares of land. This level of mechanization was higher than in the United Kingdom because the farm structure was different: under a regime of high support prices, and accompanying subsidized credit and guaranteed markets, there is every inducement for even the smallest farmers to mechanize production.

Second, although the legal and financial inducements to farm amalgamation are few within the European Community, the economic gains from larger units, from 'shedding' labour and acquiring ever more capital equipment, are considerable. The pressures on the production margins of small farmers, even those with milking machines and tractors, are further accentuated by the concentration of farm ownership. One effect of agricultural support and the technological treadmill is, therefore, greater specialization. It is in the automatic consequence of inducing investment, and losing employment, in farming.

ENVIRONMENTAL MANAGEMENT

André Gorz makes a distinction in his work between what he calls 'environmentalism' and 'ecologism'. Environmentalism refers to planned intervention to secure improvements in environmental quality, much of it undertaken by the state in developed countries. Ecologism represents much more deep-seated objections to the way the environment is handled in such societies: 'Ecologism is not content with correcting the environmental impact of the system. It examines the reasons *why* certain technologies and forms of production and consumption were promoted in the first place'.

In the developed countries, as we have seen, the state already intervenes to correct or modify the full effects of the development process on the environment. In such countries, the 'core countries' for the modern food system, the popular mass diet is already well established, and agriculture employs only a small proportion of the population.

Under these circumstances, managing the wider environmental implications of the food system assumes importance largely because basic needs have been met. Environmental policy interventions usually take two forms as far as food system activities are concerned: the protection of 'valued' environments from the pressures of human economic activity and the reduction of market pressures on specified groups of agricultural producers, to enable them to act more 'sustainably'.

Both kinds of intervention are, arguably, holding devices, since they can be continuously negotiated and renegotiated. In fact the more interventionist the state becomes the larger and more vocal the environmental lobby usually becomes, exposing new environmental risks and dangers, to which it is claimed government should be responsive. The principal objectives of environmental policy, in relation to agriculture, speak to the priorities and concerns of urban middle-class groups: amenity recreation, species preservation and aesthetics. The membership of environmentalist organizations has increased rapidly in recent years. In developed countries the existence of protected areas does not diminish opposition to environmental management decisions, but it does draw attention to some of the problems and limitations of what we have described as the 'modern food system' model.

In developing countries the situation is usually markedly different. Environmental conflicts in developing countries are about poor peoples' access to the resource base in the struggle to maintain livelihoods, and the conflicts which are associated with changes in resource use in the course of economic development.

Activity	Percentage of people (aged 12 or over) taking part in each activity during 1984
Drives, outings, picnics	55
Visits to historic buildings	39
Long walks	33
Visits to country parks	27
'Pick your own' (PYO) farms	23
Nature reserves	13

Countryside Commission, National Countryside Recreation Survey, *SocialTrends* 17

Note: In any four week period during 1984, 40 per cent of the population made no visits to the countryside, while 17 per cent made nine visits or more.

Membership of Environmental Organizations

Organization	1970	1975	1980	1985	1990
National Wildlife Federation	2,600	n.a.	4,600	4,500	5,800
Sierra Club	114	153	182	363	566
National Audubon Society	105	255	310	425	515
Wilderness Society	66	n.a.	63	97	363
Environmental Defence Fund	10	40	45	50	150
Greenpeace	"	6	80	450	2,000
National Resources Defence Council	"	15	35	65	140

'Sustainability' in this context is not primarily a consumption goal of the relatively leisured middle class, but a concept which, if it is espoused by anyone, is espoused by the rural poor themselves and usually within the context of human rights. Of course, this is not always the case, and the fact that the rural poor often stand in an ambivalent position to more sustainable systems of agricultural production is evidence of the difficulty in securing livelihood advances.

It is worth emphasizing, too, that the interest of people in developing countries in their own environments does not put a seal on environmental conflict in these societies. There are also issues of the 'international commons': common property resources, as well as open access resources to which no property rights apply. Many of the world's life-support functions are dependent on the protection of resources in large tracts of the South; genetic materials provide the best example, as we have seen, but increasingly even issues like rain forests, climatic change and the disposal of toxic wastes involve relations between the developed and developing countries.

The globalization of the modern food system ensures that changes in resource use, and the environmental effects of new technologies, are transmitted from one geographical location to another, just as patterns of food consumption have become increasingly international. Policy and management interventions to secure more 'sustainable systems' cannot be judged simply in terms of their own logic and internal dynamic; they need to be considered within the broader context of production, consumption and the transformation of resources which defines the modern international food system.

As the British response to the World Conservation Strategy made clear, 'any useful analysis of the reasons for natural reserve destruction would have to include questions of wealth distribution *within* and *between* nations'. The point about environmental management, as it is practised in the developed and developing countries alike, is that it takes as a 'given' the distributive consequences which the market produces in the course of development. In developed countries, as we have seen, increasingly vocal and well-organized groups have mobilized around environmental objectives. In developing countries only the poorest and most marginal are likely to do so: rubber tappers in the Brazilian Amazon, Chipko activists in India, and women farmers in the Kenyan 'Green Belt' movement. The politics of environmental movements, as well as management interventions, have a different basis in the South:

> The more aid flows towards those sectors in developing countries favoured by political, industrial and commercial criteria, the less likely they are to arrest such environmental damage as is occuring. This is because the greatest environmental damage, and needs, tend to occur in the rural areas and amongst the urban poor of the least developed countries (with the obvious exception of industrial pollution which is not insignificant in some developing countries). In the developed countries the management of the rural environment is contested by two opposing interests: those of farmers and conservation bodies. The environmental lobby has assumed increasing importance in the United Kingdom

because environmental values are located in both resource-rich and resource-poor areas. Although some relatively remote areas are under populated and offer few attractions to agriculture, most rural areas are within reach of heavily populated urban settlements and land uses are strongly contested.

In North America, Australia and parts of Europe the environment presents itself in another form, that of wilderness, which has served as a focus for the environmental lobby, both as intellectual inspiration ('deep ecology') and as a barometer of cultural, and hence political, concern. Nash, in *Wilderness and the American Mind,* called these areas of wilderness 'institutional containers', vital because they have 'resisted' civilization.

The juxtaposition of dense urban populations and resources worthy of conservation, in the developed countries, enhanced the importance of 'protected area' status, such as that of National Parks. Most of these areas had little agricultural potential, but enormous cultural and political weight was attached to their conservation.

The difficulties afflicting areas of some agricultural importance, which contained or bordered zones of environmental value, were rather different. Strategies have had to be evolved for reducing productivist pressures in these areas, either through 'setting aside' land deemed surplus to agricultural requirements or by shifting towards 'lower-input' farming. Beginning with physical planning controls, and protection of areas designated as being of special interest, the move within the countries of the European Community has been towards more economic incentives to practice 'sustainable agriculture', including measures to support farm families as well as farmed landscapes. The motivation to undertake more management interventions has been provided not only by heightened environmental concern, but the onerous budgetary costs of supporting agriculture and production of food surpluses.

The comparison with the developing countries makes it clear that environmental management principles cannot simply be transferred from North to South. In developing countries the contest over natural resources, their use and ownership, is between the rich and the poor in rural areas. In most cases environmental agencies are influenced by external pressures—such as those associated with opposition to tropical deforestation—and, in any case, such agencies have no 'teeth'. Protected areas such as Biosphere Reserves are demarcated, but the means with which to ensure their protection are rarely available; budgets are minimal, 'policing' activities random or unenforceable.

In areas other than protected areas there is little, if any, outside intervention to protect the environment. In some cases, agricultural technology is disseminated which is more 'environment-friendly', such as that of Farming Systems Research (FSR), but this technology is encouraged largely because high-input agriculture is impossible on marginal or resource-poor land. Only relatively recently have 'sustainable' objectives begun to be incorporated in production programmes for small farmers, but the vast majority of the rural population, particularly rural wage labourers, lie outside the reach of existing 'sustainable' policy packages. It is extremely unlikely that policies like the 'polluter pays' principle could be incorporated within the gamut of policies influencing rural resource uses in developing countries.

The pressures mounted within the international community to call a halt to environmental degradation in the South have had little impact on the economic policies and practices that largely govern North-South relations. One much-debated exception is that of 'debt-for-nature' swaps, in which some areas of tropical forests are 'protected' from development, and their value written-off against part of the country's external debt. In practice, however, such swaps feature in only tiny areas of endangered land, it is unlikely that they can be adequately protected, and in any case the decision to do so was not arrived at democratically but by administrative fiat.

We have yet to see even the beginning of effective local participation, by indigeous communites and others, in the long-term sustainable management of tropical forest areas.

What has emerged from the 'debt-for-nature' debate is a measure of hostility between some governments in the South, anxious to protect their sovereignty and forced to increase foreign exchange earnings by selling products that carry environmental costs, and what are sometimes perceived as 'interfering' First World environmentalists. Developing country governments argue, not unreasonably, that they cannot be expected to practise sustainable development when their paymasters, the developed countries, fail to do so.

AGRICULTURAL INTENSIFICATION

The productivist model, which lies at the heart of the modern food system, has both direct and indirect environmental effects-environmental problems in the immediate locality, and others at one remove. The environmental problems associated with agricultural intensification are, in the first instance, experienced as a direct effect of the implementation of the agricultural technology/policy model. As we shall see, these first-order problems hold for both developed industrial societies and for areas of modern agriculture within developing countries.

Developed Countries

The effects of agricultural intensification can be illustrated by taking as an example the removal of hedgerows. Similar accounts could be given of the effects of drainage, or the application of chemicals or intensive livestock production.

Most factors of production are in relatively elastic supply, but land is the exception. The supply of land can normally only be increased through 'improvement' or intensification. As agriculture feels the effect of greater capital investment, prices increase and these become capitalized in land. One effect of this process is to increase the opportunity cost of unproductive land, such as hedges and field margins.

The farmer will view the conservation of hedgerows as an opportunity foregone—he has forfeited the opportunity to increase his income by producing more food. This tendency to regard conservation activities as a cost to current income is exaggerated because, under the system of grants to defray capital costs within the European Community, farmers are induced to specialize, substituting capital for labour in many of their operations.

As Cheshire argues, 'while labourers could milk cows, tend the sheep and with a simple binder, harvest the corn, milking parlours can only milk cows, and combine harvesters can only harvest cereals'. By increasing output the farmer was induced to replace organic with chemical fertilizers, and to use chemical sprays for pest control. Both sprays and fertilizers depend on subsidized machinery.

Under specialized production hedgerows were unnecessary, since they served no purpose as livestock barriers, and animals had been brought in from the fields. The use of larger machinery also meant that fields needed to be larger, persuading farmers to recoup their increased capital costs by removing unproductive hedges and ploughing over footpaths, which interfered with the movement of their mechanical equipment. The environmental problems that have accompanied intensification are the 'effect of a system which systematically establishes financial inducements to erode the countryside, offers no rewards to offset market failures and increases the penalties imposed...on farmers who may want to farm in a way which enhances and enriches the rural environment' (. The example of hedgerow removal is just one of many similar examples of foregone production benefits being converted into environmental losses.

Among the most important effects of intensification is the use of nitrogenous and other chemical fertilizers. Within the European Community the use of chemical fertilizers has assumed increasing importance. According to the FAO, consumption of nitrogenous fertilizers in the Netherlands is 558 kg per hectare, Switzerland 436 kg, West Germany 427 kg and the United Kingdom 356 kg. There is evidence that these levels of use leave significant residues in the soil.

A report on Paris Basin agriculture shows a nitrogen surplus of 240 kg per hectare over a seven-year period. Eutrophication of lakes, rivers and coastlines caused by excessive nutrient loading from nitrate and phosphate pollution is also widespread. The FAO notes that recent acute deterioration of marine ecosystems in the Kattegat and the Adriatic Sea, for example, has had a significant impact on fisheries and tourism in these areas.

The intensive production of animals in feedlots has had similarly important environmental effects. It is estimated that each year British farmers dispose of 200 million tons of animal manure through the river system. The disposal of slurry is now a major environmental problem. A survey by the Water Authorities Association in July 1986 revealed that incidents of silage effluent pollution were almost double those of two years earlier. In 1985 Anglia Water reported that half of all fish deaths leading to prosecutions, and 60 per cent of its pollution prosecutions, were attributable to slurry disposal. The excessive use of animal slurry can lead to concentrations of up to 10 kg of nitrogen per hectare. The mineralization of the soil from increased large-scale nitrate leaching can pass into both surface and groundwater. Already, in the Netherlands, the nitrate content of shallow sources of fresh water is too high to make it usable.

The average nitrate content of aquifers in the Paris Basin has doubled from 20 to 40 mg per litre during the last twenty years, while in some areas of France, such as Beauce and Sologne, the nitrate content of groundwater exceeds 100 mg per litre. Studies of more than 6,000 wells in North Rhine-Westphalia showed that two-thirds had nitrate contents of 50 mg per litre or above, the maximum permitted under EC Directive (80/778, 15 July 1980). A third of the wells exceeded the very lenient limit of 90 mg per litre permitted in West Germany.

The cost of reducing nitrate concentration in groundwater to within acceptable limits in West Germany has been calculated as 2000 DM per hectare. Facts such as these have led to alarm within EC policy circles. European Commission guidelines state that: 'indefinite exemption from the principle that the "polluter pays" can hardly be given to agricultural holdings, least of all those which, in terms of the way they are structured and operated, are to all intents and purposes industrial undertakings'.

The conversion of forests, pastures and wetlands to other uses has led to environmental losses in other directions, notably the loss of wildlife habitats, hydrological changes and increased risk of erosion. Land reclamation practices have brought about radical changes in the environment during the last decades and, although the rate of land conversion is decreasing within the European Community, it is still a contributory factor in further environmental deterioration.

In the United Kingdom, for example, a dramatic loss of valuable biotopes and wildlife habitats has occurred since 1945. Almost all unimproved natural grasslands, including herb-rich hay meadows, have disappeared, as have 40 per cent of lowland acidic heaths, 30-50 per cent of lowland woods and 30 per cent of unimproved upland grasslands, heaths and blanket bogs. As the FAO report to the United Nations Economic Commission for Europe expressed it, 'the combined effects of modification of traditional agricultural landscapes, increased use of pesticides and intensification of agriculture have reduced wild species abundance and diversity in most agricultural areas in Europe'.

There are many interrelated reasons for loss of soil productivity which result from agricultural intensification and heavy reliance on external inputs. These effects are not confined to developed countries, like those of Western Europe. The more or less proximate effects of agricultural intensification are also felt in areas of the Third World where modern agriculture has become established, often with fewer controls on adverse environmental effects than exist in the developed countries.

These areas, which we can term 'enhancement zones', perform a critical role in food production in the South, and particularly the production of food staples such as rice, which are grown under highly intensive irrigated conditions. Through which modern intensive agriculture has served to transform the environment of developing countries, suggest that the modern food system does not need to be fully elaborated within developing countries, from both a consumption and production standpoint, for new agricultural technologies to carry important social and environmental implications there.

In the developed countries the consequences of the modern food system have been challenged on the grounds that financially underwriting modern agriculture has proved extremely expensive: the costs of agricultural support are paid by the taxpayers and consumers within the European Community, whose interests are now beginning to be championed by political leaders as well as lobbies. The other major objection to agricultural intensification has come from the environmental groups opposed to high-input farming, or interested in preserving the amenity and recreational aspects of the countryside.

Government support for agriculture in the developed countries has been financed from domestic budgets, and the financial costs have been passed on to the population of these countries. In the developing countries the circumstances are rather different: new developments in technology and the implementation of agricultural development policies, have been financed through external loans and credits. For example, the International Agricultural Research Centres, which form part of the CGIAR system discussed in the last chapter, are largely supported by individual governments in the developed world, development agencies such as the World Bank and by large foundations such as Rockefeller and Ford.

To the extent that these support costs are visible, and many of them are not, it is difficult for the people of Third World countries to mount an effective opposition to the way funds are disbursed on their behalf. Similarly, people throughout the developing world are often very aware of the environmental costs of modern agriculture in areas of intensive production and of the environmental costs in 'low resource areas' which have been increasingly marginalized from the development process.

However, as we shall see, in most developing countries it is much more difficult to establish environmental priorities in the face of the 'gains' expected from economic growth, including gains in agricultural production. Yet the heavy cost of environmental losses should not be underestimated, as we have seen. Developing countries are effectively caught in a 'scissors movement'; the costs of conserving natural resources are borne by them, while they also bear the cost of revamping their economic policies under the exigencies of structural adjustment, and other packages emanating from international financial institutions.

In many parts of the Third World, especially Asia, the Green Revolution has often been depicted as a success of 'science' and 'technology' in overcoming the obstacles to increased agricultural production. In fact, as a number of writers have indicated the most consistent factor in the production successes attributed to the Green Revolution has been public sector investment, often reinforced by externally-secured finance, and including a series of support measures affecting prices, marketing arrangements and crop procurement, the transfer of technology and input manufacture. The effects of the modern food system have been felt, if not directly then indirectly, in the way developing countries have sought to subsidize the adoption of increasingly intensive agricultural packages.

The 'successes', in production terms, of the Green Revolution in most areas where wheat and rice are staple crops can be attributed in large measure to the transfer of the agricultural technology/policy model first developed, in its modern form, in the United States in the 1950s and 1960s. In the developing countries the International Agricultural Research Centres, in particular, under which research institutions in the United States developed and transferred new technologies from the laboratory to the land.

The adoption of these technologies, in the developing countries as well as in the North, had more to do with the incentive structure that surrounded both their use on farms and the research and development exercise. One important aspect of the expansion of the modern food system is that components of this system, if rarely the system in its entirety, have served to transform the context of agriculture in so many Third World countries, despite structural constraints to the adoption of the fully elaborated industrial food system. Essentially, the adoption of 'high' capital-intensive technology under conditions in which most of the rural population live in material poverty, has in many cases served to exacerbate inequalities within regions.

Under these circumstances, ecological problems, especially increasing fragility of the ecological system, are closely related to a reduction in livelihood choices for poor people, and their efforts to re-establish a viable livelihood are frequently costly in environmental terms, whether in resource poor areas or in areas of capital-intensive agriculture.

Recent papers from the World Bank have suggested some useful policy options, which would enlarge the management interventions available to government. One example is the removal of pesticide subsidies, which are costly in economic terms, for most developing countries, and undermine the stability of agriculture. The removal of these subsidies, or the introduction of a pesticide tax, would reduce pesticide consumption, and give farmers an increased incentive to adopt more sustainable, biological alternatives (such as those indicated under the general rubric of Integrated Pest Management). In the case of Indonesia, it has already been demonstrated that reducing pesticide subsidies brought fiscal savings of US$150 million annually over a period of four years. As World Bank economists have shown, the reduction of these subsidies contributed to more pest-resistant rice varieties becoming available, and Integrated Pest Management practices became more widespread.

Marginal Lands

The secondary, or indirect, effects of the food system on the environment in both developed and developing countries are more difficult to delineate. The process through which the modern food system becomes established at the farm level—concentration of holdings, specialization and intensification of production—carries implications for areas other than those of modern agriculture. The cost of achieving gains in productivity in one locality is often passed on to others; through environmental costs or 'externalities'. As we have seen, intensification needs to be understood not simply in terms of the technology employed but also in terms of a 'package', comprising guaranteed prices, subsidized chemical and mechanical inputs and government procurement after the harvest.

In some developed countries, such as those of the European Community, less advantaged regions receive subsidies from central government precisely because they are adversely affected in social terms—they are less competitive in terms of their economic performance. In developing countries, however, areas where natural resources are poor, especially semi-arid regions, are likely to fall completely outside the ambit of public policy.

This neglect serves to accentuate the differences, the unevenness of the development process. It is more difficult for small farmers, for example, to gain government support for their farming system; they are left to manage an inadequate technology, even if this technology is environmentally sustainable, ill-adapted for competition with the technology employed in more favoured regions.

Many households, in marginal areas of Europe, Australia and North America, such as the Appalachian Mountains, as well as the Third World, have to supplement the income they obtain from their land by other forms of livelihood. They become part-time farmers, working in the city or in tourism, or migrating on a seasonal basis to other regions of the country. Living in areas that are already poor in terms of productive resources, but frequently rich in terms of natural species or aesthetic quality, they are powerless to do anything in the face of the differentiating effects of modern intensive agriculture.

This chapter is concerned with both the immediate and the more indirect environmental implications of the modern food system. It needs to be emphasized that a series of 'third-order problems' also exist which we are only beginning to delineate: that is, problems such as those arising from the introduction of genetically engineered species into the environment. The discussion of biotechnology and genetic engineering touched on these issues, but it is too early to elaborate at length on the social implications. These 'third-order problems' are likely to grow in urgency and include not only the compatability of 'new' life forms and existing ecosystems, but also the relationship between conserving species' diversity in laboratories, and gene banks, as well as in the wild.

The Management Response

The early part of this chapter addressed the question of agricultural policy instruments, and the way these had been used to support intensification, at the cost of environmental considerations. Given the high visibility of environmental issues and, in the developed countries, the acknowledgement that environmental considerations require changes in agricultural policy, it has become important to review the various ways in which the rural environment can be more effectively 'managed'. Beginning with an analysis of environmental management in rural Britain, and extending this to the rest of the Community, this section goes on to assess the kinds of farm-level response which have been induced by recent policy making. The question posed by this discussion of sectoral and farm-level decision-making, is whether it carries implications for developing countries, where the penetration of the modern food system is also eliciting environmental management interventions. This means looking very carefully at the assumptions underlying environmental management itself.

As Potter observes, the increasing concern with long term support for farming in Britain and the European Community generally has 'coincided', especially in the British case, with public concern about the environment and its conservation. As we have seen, policies which encourage agricultural intensification, specialization and the concentration of farm ownership in fewer hands have been called into question by the existence of farm surpluses and the heavy financial cost of supporting the agricultural sector. It needs to be emphasized, however, that it is public involvement in conservation, and the lobbying of environmental interest groups in Britain, which has called the model itself into question.

The agriculture and environment debate in the United Kingdom dates back to the early 1960s, although it did not reach the European stage until almost twenty years later. Beginning in the 1960s the public's attention was alerted to the cumulative effects of using pesticides in agriculture. The Nature Conservancy began to reveal that there had been dramatic losses in some wildlife species, such as peregrine falcons, due to the use of organo-chlorines. Within a decade this concern, fostered by the growth of the countryside lobby, had extended to the loss of whole habitats, such as moorland, heath-land and ancient woodland, losses which were attributed to agricultural intensification and land reclamation.

Groups such as the Council for the Protection of Rural England (CPRE) which had considerable following, especially in the more scenically attractive parts of the British countryside, drew attention to these environmental losses. Other controversies were more long-standing but no less vituperative: the ploughing of moorland in the Exmoor National Park, for example, drew attention to the limited powers exercised over supposedly 'protected areas'. Land use controversies began to engage public attention as much as environmental pollution or contamination from intensive farming practices. Furthermore, the effect of EC membership in accelerating these losses gave rise to additional concern at the environmental costs of 'productivist' agricultural policies.

The loss of the countryside's amenity value was a particularly British concern. As Potter shows, the agenda was partly set by relatively autonomous bodies known as 'quangos' within the structure of the British government, notably the Countryside Commission and the Nature Conservancy Council. In 1974 the Countryside Commission published its 'New Agricultural Landscapes' study, which documented the ways in which familiar features of the British countryside, such as hedges, copses and ponds, had been removed, as farmers deserted mixed husbandry for more specialized farming operations.

By 1977 the Nature Conservancy Council (NCC), largely staffed by trained environmental scientists, with no obvious 'political axe' to grind, had reported that existing policies, which sought to protect sites of special environmental interest rather than more integrated land use practices, were seriously deficient. By 1984 the NCC had revealed that only 3 per cent of lowland grassland had remained unaffected by agricultural intensification, since 1945, and that ancient lowland woodlands had been reduced by between 30 and 50 per cent in the same period.

While the environmental lobby was beginning to raise its flag during the 1960s and 1970s, evidence was emerging that the agricultural lobby was being forced to adapt, slowly and painfully, to the realization that it could no longer count on the unqualified support of British governments. Two agricultural White Papers in the 1970s departed, in significant ways, from the conventional wisdom that, as the major land-using industry, agriculture could expect to expand unchecked. Although no revised national policy existed which integrated the agricultural restructuring being urged on the European Community with more attention to environmental objectives, questions such as the appropriate size of the agricultural industry in terms of food and fibre output had become matters of policy debate. As a recent research paper makes clear, a subtle shift had occurred between 1975 and 1979, in which the Ministry of Agriculture conceded that the 'national interest' was distinguishable from that of farmers.

By 1981 the tide of concern over the demise of the British countryside led a Conservative government to introduce a Wildlife and Countryside Act which provided a legal basis for management agreements between farmers and the NCC on farmed land of special scientific interest. This Act allowed for payments to be made to farmers, by way of compensation for foregoing modern agricultural practices, in return for agreed conservation practices. Owners and landowners were required to seek permission from the NCC if they intended to change their farming practices.

The scheme attracted considerable criticism for being 'too little, too late', resting on the farmer's goodwill, and proving very costly to the Exchequer. Although the shift away from unbridled productivism had occurred, the Government seemed anxious to proceed by agreement and co-operation with farmers, almost whatever the cost. Voluntary and informal structures of land use planning and regulation were very much the method favoured by the Government.

It is this concern with 'management agreements' that marks out recent policy interventions in the industrialized countries. According to a recent report from the OECD there are three desirable aspects of such agreements. First, the environmental benefits have to be realised effectively. For example, if the aim is to prevent cropping, then the cropping rights should be permanently acquired under the agreement. Second, the payment made in compensation for the reduction in expected net farm income, which the adoption of new management practices is likely to produce, should not exceed this 'opportunity cost'. Third, the agreements should not be in force long enough to enable farmers to bargain about the level of compensation, by refusing to operate under the new conditions until all their demands are met. It will be clear from these 'desirable conditions' that it would be impracticable to introduce similar agreements on a large scale in most developing countries.

The mounting controversy about agricultural support, however, did begin to alter the policy debate by linking the redirection of agricultural support to rural conservation measures. Within the EC structures legislation and the enabling legislation contained in the 1986 Agriculture Act, financial aid began to be offered to appropriate farming practices as well as to farmers who refrained from using inappropriate practices. A number of areas considered 'environmentally sensitive' were outlined in the 1986 Act, and about 750,000 hectares of land were designated in the United Kingdom.

Farmers were being paid, for the first time, to 'produce' countryside, something which, they did without significant agricultural support in the era before agricultural intensification. The principle that the interests of farmers were distinguishable from those of the British countryside was conceded in the 1986 Agriculture Act, which 'imposed a statutory duty on Agriculture Ministers to balance the conservation and promotion of the enjoyment of the countryside, the support of a stable and efficient agricultural industry and the economic and social interest of rural areas'.

Although the agricultural lobby no longer commands undivided support, even from Conservative governments in Britain, and the environment lobby has increased in size and authority, the mechanisms through which agricultural support can be shifted towards better conservation are still in their infancy. Writing about the United States, where the conflict is more often between agricultural intensification and sustainable agriculture, rather than'countryside' with amenity and recreational value, Buttel argues that farmers are most unlikely to change their practices to meet environmental guidelines: State policy-makers have been unable to call upon farmers to collectively and voluntarily reduce their production during phases of overproduction. Given the inability of farmers to control their production, state action to reduce production by rationing land resources has been undertaken because rationing other inputs (ie. petrochemicals) would be strongly resisted by farmers and by agribusiness firms.

Some commentators, such as Potter have argued that the preservation of existing farms is a necessary component of any wider conservation goals in European rural areas, and that the offering of income aids to farmers who agree to make their farming practices more environmentally sensitive is the most direct and powerful way of advancing future rural environmental policy. However, the same writer is equally aware of the limitations of such a small-farm approach.

He writes that: realistically, though, there are limits to the extent to which a Farm Survival Policy can be orientated towards conservation objectives. With the notable exception of environmentally sensitive areas it is unlikely that the disadvantaged and intermediate farms that will be targeted in future will be located in areas of greatest environmental potential or vulnerability. The very transparency of expenditure on income support presents another difficulty: it may not be desirable to seek to achieve long-term environmental goals under expenditure headings that may be vulnerable to political pressure.

It is not clear, then, that the situation in Europe is as strikingly different from that of North America as first appears. The powerful economic lobbies behind high-input farming have not been seriously affected by the new calls to conservation. Indeed one feature of so-called 'Green consumerism' today, is that commercial interests with dubious environmental records are often loudest in their support for more 'sustainable' practices. The political climate has changed in recent years, and this is not an unimportant fact in itself, but the model which encouraged agricultural intensification as part of the development of the modern food system, has been dented rather than overturned. In the process, 'management' has acquired both greater credibility and a more uncertain future.

MOVEMENTS AND SUSTAINABLE DEVELOPMENT

Most of the recent discussion of 'environmental management' in developing countries has paid scant attention to the role of local people. Environmental management is defined almost exclusively terms of a rigidly 'top-down' approach. It is assumed that by incorporating better systems of environmental accounting in the national accounts, and by ensuring that the negative costs of development are allowed for within project budgets, progress will be made towards better management of the environment.

Desirable as it is that economics takes environmental problems seriously, such measures are unlikely to affect the balance of forces which cause poor rural people to place excessive strain on their environments. The 'official' discussion of environmental management in the South makes barely a reference to the increasingly important experiences, and literature, on environmental movements there. The remaining section of this chapter focuses attention upon the need to re-examine our knowledge and understanding of the environment, in the light of the experiences of other cultures far removed, socially and geographically, from the 'core' of the modern food system. It is clear that rural development projects have often neglected local, ecological and cultural factors in the South. It is also clear that structural factors—which influence people in not acting 'sustainably'—have frequently been ignored. We need, in fact, to take account of both the cultural and structural dimensions in any definition of sustainability. Then we can 'begin to ask whether there are patterns of myth and social organisation, which are more likely to ensure and maintain man's *[sic]* survival base'. It is clear that environmental problems are not new. Historical, and archaeological, evidence suggests that early societies often behaved in ways which were damaging to their environment.

These societies also had to 'solve the basic problem of producing food surpluses and collecting raw materials from rural areas to sustain large urban populations', a problem that faces countries in the South today. However, such societies, unlike the South today, were not faced by twin assaults from rapid economic and demographic growth, on such a scale. They did not exist within a global context which placed their own food systems and values in jeopardy.

If we are to understand fully the relationship between sustainability and development, we must begin by recognizing that they are paradigms with historical as well as geographical antecedents which need to be located within specific social and economic contexts. The point at issue is that societies differ in the way they prescribe courses of action towards the environment. Too often we assume systems possess universalistic characteristics, including the food system.

Such universalistic approaches to social and political systems often leave no room for human agency within the framework of explanation. The question we might address in exploring environmental movements was posed by Giddens in the following way: how does the 'practical consciousness' that people acquire from their society, and from their environment, relate to the 'structural properties of the system'? Environmental movements include movements of resistance to the encompassing food system. Under what circumstances are the structural properties of such a system called into question? To answer this question we might begin, not with an analysis of political conflict over the environment, but with the way in which the 'environment' is understood. Instead of working 'backwards' from social movements to consciousness, we can work 'forwards', from the formation of environmental consciousness to the organization of alternative forms of social action to redress environmental problems.

It is clear that the way in which knowledge is gained about the environment and the way in which this knowledge is used have considerable bearing on the question of human agency, and with it environmental management practices. Howes has argued that in agricultural systems engaged in monocrop cultivation the world view is one in which 'society is conceived as an entity apart from Nature'. It is clear, too, that many non-Western cultures possessed 'scientific theory' before Western science, as we know it, had fully developed. Looking at examples from South Asia, Goonatilake claims that, 'their large-order cognitive maps anticipated the non-manipulable, and distant reality of modern science'.

Furthermore, these kinds of knowledge system often codify knowledge in ways that ensure its cultural survival, long after the reasons for certain observances have passed. In some cases 'after an initial map has been made and legitimized...it tends to continue to be believed for considerable periods of time'. A useful distinction, although it should be emphasized that it exists *within* societies as well as between them, is between what Feyerabend calls *historical traditions* and *abstract traditions.* The two traditions can coexist in a society at any point in time, but some cultures give more emphasis to one tradition than to the other. Historical traditions involve a different discourse from abstract traditions. It is assumed that 'the objects already have a language of their own' and it is the task of people within a society to learn this language. Their task is to 'learn the language of the objects as they are, and not as they appear, after they have been subjected to standardizing procedures' and experimentation. According to this tradition there is no such thing as pursuing 'objective truth', since truth involves both an object and a subject. Feyerabend compares this with what he terms an abstract tradition.

Abstract traditions are different in that they involve framing statements as a means of building up a model of what happens in nature. According to Feyerabend, 'the statements are subjected to certain rules (of logic, testing and argument) and events affect the statements only in accordance with the rules' It is possible, then, to make scientific statements without personal contact with the objects being described. Both these epistemological traditions exist, at different levels, in most societies, but within the Western scientific tradition much more emphasis is given to abstract traditions.

In this respect the development of Western science is best understood in the context of Judaeo-Christian thinking, which has often appeared inimical to 'scientific' method. For example, Goonatilake argues that religious and philosophical objections to scientific method in Europe would not have occurred in South Asia, where most religious beliefs emphasize the role of both abstract and historical traditions in arriving at valid statements about the world around us. The first point that needs to be emphasized in establishing a framework for understanding environmental conflict in the South is that a society's epistemological framework will influence its understanding of the environment. What we perceive as 'environmental' problems may be interpreted in other cultures as essentially due to the breakdown of social and cultural practices and beliefs.

In addressing these questions we need to be aware that our understanding of Societies very unlike those of the industrialized countries frequently distorts our understanding of their environments. By the same token, the rediscovery that knowledge-systems other than our own have importance for understanding our relationship to the environment has become a central, but contentious, tenet of recent Green thinking. As Luke observes, some Green thinkers have misunderstood the importance of non-Western science. 'Deep ecologists' have sometimes equated small-scale, non-Western societies with 'primal societies', implying that 'primitives' provided an echo of our own pre-Enlightenment past.

This rather naïve view of the superiority of non-Western epistemologies fails to recognize that myths, magic and ritual are 'the functional equivalents (and perhaps the conceptual antecedents) of Enlightenment science and technology'. This kind of simplification and misunderstanding also errs in lifting ecophilosophies out of their cultural context—Taoism or Buddhism—'with little consideration of any cultural grounding', when it is the specific cultural context of environmental epistemologies that gives them their force and vitality. As an approach, the close association of indigenous knowledge with 'answers' to current environmental problems is frequently ethnocentric, a historical and reductionist, substituting descriptions of the world for more rigorous and less ethnocentric analysis.

Other examples exist of societies closer to our own in which environmental problems are constructed from very different political and social milieux. In Eastern Europe today, for example, the commitment to radical political ecology, and the growth of Green movements, is also linked to processes that are not strictly 'environmental' in the narrow sense employed in the West Conflicts over ethnic identity, and the struggles for civil and human rights in contemporary Eastern Europe, have incorporated—and adapted to—environmental demands, and often express these demands in language drawn from the discourse.

The second, important dimension of relations between the environment and society is the economic dimension. The environment is subjected to structural changes over time, linked to economic growth and the development process. If we confine ourselves to the relationship between agriculture and the environment we can identify several features of the development process which have major implications for environmental systems, among them: the production of food as a commodity rather than for household consumption; the adoption of 'high-technology' energy systems on the farm, and in the food processing industries, rather than 'low-technology' energy inputs; and the role of technology in reducing the pressure exercised by population increase on limited natural resources.

These factors will be looked at in turn. In some societies agricultural practices underwrite sustainability, in the sense that consideration of the long-term implications of current agricultural practices is enshrined in farm management practices. For example, Wilken shows how some small farmers in Central America, without recourse to fossil fuels, effectively 'bank' their own labour, and that of their household, investing labour in resource management practices that make sense within a longer time-frame. They are seeking to reduce future operational costs on the farm by making major current investments in labour-intensive management now.

As we have argued above, in some societies such practices become ritually encoded. Rappaport shows how, under stable conditions, the ritual enactment of myths may not only help to maintain the social cohesion of the group (a point often made by anthropologists) but also help to sustain the natural environment. Of course, the value placed on a resource is linked to other social referents, such as the concept of ownership, as well as economic factors like the cost of extraction.

The development of wider markets for food products, and the adoption of production techniques that depend on sophisticated off-farm processing and industrial inputs, does not necessarily mark the point at which food becomes a commodity—in most societies food has been a commodity for a long time. However, it does mark the point at which the conversion of food from a use value to an exchange value has the greatest impact on the way the environment is managed, and ultimately socially constructed.

Food is taken from the province of reproduction in small-scale societies and transferred to that of production and consumption in more complex societies. Eventually, in what we can term 'post-industrial societies', the locus of dissent shifts again, this time from production relations to those of consumption. Some of the most important implications for the evolution of the food system are represented by resistance to the introduction of new, processed 'convenience' foods. The food system also affords an excellent example of the way that energy use is linked to the capacity of ecosystems to retain, as well as transfer, value. The process of development, as we have seen, is usually considered in relation to the movement of capital and labour, rather than the exploitation of natural resources. The structural and spatial aspects of transfers of value from one part of the (developing) world to another part of the developed world have been emphasized by O'Connor and Smith. Without dissenting from the significance of these forms of surplus extraction, it is important to emphasize that transfers of energy from within one system to another can also be seen as transfers of value.

Bunker uses both ecological principles, and the laws of thermodynamics, to assert that 'our calculus of value must include not just the labour and capital incorporated into commodities, but all forms of energy and matter'. In his view, 'the embodiment of energy in economic and social organization encompasses far more of the essential differences, and relations, between core and periphery, than measures linked to commodity production and exchange can'.

Economics has also experienced considerable difficulty recently in incorporating the full implications of changes in environmental systems within its analytical framework. The Malthusian tradition emphasized that population pressure placed limits on the capacity of natural resources to feed the population. In fact, like many economists, Malthus did not have a great deal to say about the environment. Barbier reminds us that the pessimism of the Malthusian tradition was only matched by the optimism of the Ricardian view of scarcity. This placed emphasis on 'relative scarcity' rather than 'absolute scarcity'; as resources became used up substitutes were found for them, extending the time-frame in which the environment could facilitate economic growth.

The problem of carrying capacity was not solved but it was 'put back', and technology was the major factor in effecting this change. Later neo-classical thinking, represented by Alfred Marshall, took an even more optimistic position than Ricardo; no tendency was predicted for profits, wages and rent to decline with increased scarcity of natural resources. The implication was that we were entering an 'age of substitution', the environment could be exploited without degradation being inevitable, indeed the scarcity of non-renewable resources might attract a higher price, and therefore lead to their better conservation. This view of 'substitution' still plays a large part in the way we conceptualize the environment today, and has been given added impetus by developments in biotechnology.

The importance of the economic dimension is that it necessarily involves considering the conversion of natural capital into social capital, economic processes of growth and development do not simply modify, they actually transform the environment.

In the course of these transformations the development process 'hands back' to nature some serious problems, including that of sustaining economic levels of production and consumption. The development of the modern food system has brought new urgency to another important dimension: the political dimension. The wider parameters of the food system have introduced this dimension into social relations on the farm, relations which were previously mediated by more endogenous factors. Linkages between agricultural producers, suppliers and markets have necessitated more joint decision making, and have led farmers and the food industry into political alliances.

As producers and consumers have become incorporated into wider market systems, their response to their tacit incorporation has often taken the form of resistance, in which class, gender or ethnic subordination is contested.

The 'fields of resistance' that result have often served as rallying points for disaffected groups with different political agendas. The concept of fields of resistance we take from Foucault. It is used to distinguish between different responses to the 'individualising techniques of power', through which systems are able to exert social control and often develop a political consensus. The individual's response to the management of this consensus lies at the heart of the resistance to industrially produced food in the developed countries. Attempts to create a consensus around modern agriculture and food policy have recently broken down.

Among the forms of resistance which environmental movements express we can include:

1. *Exploitation,* through which the individual is controlled via market forces and the production process;

2. *Domination,* through which the individual's identity is defined in terms of gender or ethnic roles; and
3. *Subjection,* where the controlling effect of dominant belief systems in the society tie the individual to an ideological position.

An analytical framework based on fields of resistance is of use in exploring the political changes that accompany the development of the modern food system, and the transformation of natural resources. For example, in the case of farmers in the United Kingdom the growth of market forces and production relations on the farm proved to be the decisive factor, at the turn of this century, in bringing about national organizations of both farmers and farm workers from different class positions. However today, farmers and landowners are united in their resistance to restructuring brought about by the market, as well as to government policies on conservation or 'countryside management'.

Farmers have become dependent on the state, and at the same time resentful of their dependence—an ambiguity behind the attraction of agrarian ideology, especially in countries like France and Germany, where farms were smaller and the dislike of state corporatism more pervasive. By contrast, the importance of *subjection,* as a field of resistance, has grown with the increasing effect of modern agricultural practices on the environment. The demand for more 'organic' food, the interest in preserving the physical condition of the countryside, and political mobilization around issues like animal rights and food quality have brought a new urgency to the interface between agriculture and the environment.

The political dimension exists in any set of environment-society relations. However, we need to ask why resistance sometimes finds expression through collective activity (land-rights, urban struggles, etc.) and why it sometimes implies withdrawing from the wider public policy sphere into an 'alternative' community or individual 'survivalist' position, at odds with the mainstream political system. These questions inevitably raise wider issues concerning the conflicting uses to which the environment can be put. The conflicts that surround resistance to 'development' bring into the open alternative strategies for utilizing local knowledge as well as conserving nature. The analysis, at this point, cannot be confined to the food system, and embraces the trajectory under which autochthonous alternatives to development diverge from mainstream economic forces under the direction of capital located in the North. These wider issues are discussed at length in Redclift.

Having considered the principal features of the modern food system from a historical and cross-disciplinary perspective, it is necessary to take the analysis forward into the realm of contemporary culture and politics. This involves analysing the growth of opposition to changes in the food we consume. This chapter has been principally concerned with exploring the effect of the developing food system on rural land use. Increasingly, the focus of concern, for interest groups, politicians and the consumer, has shifted to what we eat, and changes in our diet in turn reflect wider changes in consumption and the use of leisure. The food system assumes its most radical and yet most familiar form in the supermarkets where we in the developed countries buy most of our food. Resistance to changes in food marketing, processing and retailing have assumed a greater importance in recent years. Today they represent a veritable counter revolution.

3

Crop Production and Marketing: Materials and Methods

INTRODUCTION

Spatial Erosion Productivity Impact Calculator (Spatial-EPIC) gave a new direction to simulate crop production at regional scale based on microscopic simulation at each small piece of land in an efficient way. The result presented in this chapter is based on existing/available spatial agro-climatic dataset but information related to soil and water has not been included spatially at this stage as they are under process of development. Using Spatial-EPIC crop model observed crop yield for the period 1995-1993 are compared with yields simulated and it is found that agreement between simulated and observed crop yield shows better correlation.

GIS based modeling of an agroecosystem is expected to give a new approach in order to provide agricultural managers with a powerful tool to assess simultaneously the effect of farm practices to crop practices to crop production in addition to soil and water resources. At present, most of the crop models are location explicit information. Therefore, development of spatially or raster based biophysical crop model with go a long way in helping us to understand many intricacies of modeling of large areas at 1 to 10 km resolution.

The large-scale distribution of crops is larges determined by climate. This study gives a preliminary result of Spatial-EPIC (which is in the process in identifying weakness of GIS data interpolation capabilities. We focused our efforts mostly on climatic dataset generation for three major crops Maize-Wheat-Rice which can be further used to analyze the response of other natural systems to climate change in a later course. The scale resolution of most crop models used in simulating climate change impacts is, for all practical purposes, a single point on the earth's surface. To get regional estimates of crop responses to climate change requires either scaling up the model simulations from point estimates or scaling up the data inputs from point measurements and then performing the simulations. Both ways will result in some loss of information due to aggregation error. The effect of spatial scale of climate inputs on crop simulation is examined in this study.

The purpose of this chapter is to examine the effect of spatial aggregation (i.e, the size of the areas from which the synthesis of multiple in situ data points into a single areal estimate is made) of climate data used in crop models and find an agreement between modeled and observed crop yields. We begin with interpolation of WMO data at the degree (almost 10km on ground) grid resolution. Specifically this chapter evaluates the relative impacts of climate and crop spatially throughout the study area.

The followings are described; the region of the study, 1) the crop model used in the analysis, 2) the modeling of climate data to a climate model grid network and 3) the procedures used to compare observed and simulated yields over a continuum of spatial scales within the grid network. The area chosen for this study compasses one of the northeastern state of India namely Bihar. The area studied is located between 21° 58' to 27° 31' North latitude and 83° 19' to 88° 17' East longitude. It is overlain at 10km grid size, which cover the entire state Bihar. Three different crops were modeled for the period of 1985-1993 using (Spatial-EPIC) a crop growth model described below. In northern India there are two distinct seasons Kharif (July to October), and Rabi (October to March). Therefore the crops modeled were Kharif maize, Rabi wheat and kharif rice at every two year crop rotation. The period between March to June (locally called as Zaid season) left fallow during the simulation. This distribution mimics the predominant crops, which is being taken widely in the area.

SIMULATION MODELS

Until now there have been a lot of studies on agricultural potential productivity but to relate actual crop productivity, model based simulations are not adequate. Estimates of on-farm and off-farm productivity are being done using experimental/point based model. Biophysical spatial based model is still lacking to compute them at regional or national level. Simulation models, such as EPIC (Willimas, 1990) model used was originally point-based developed by USDA-ARS. The first results of Spatial-EPIC being developed by the author is presented in paper together with mathematical presentation of the biophysical crop simulation process. The model is designed to predict what would happen under different conditions, such as change in temperature, precipitation, atmospheric CO_2, soil moisture and so on at regional/national scale. The results presented in this chapter are based on climate data. A model developed by Richardson (1981) was selected for use in EPIC because it simulates temperature and radiation, which are mutually correlated with rainfall. The weather generator module offers a convenient method of obtaining the numerous long sequence weather data that are required for the simulate

Spatial-EPIC

Spatial-SPIC is a crop simulation model being developed to estimate the relationship between soil erosion and crop productivity, which has been implemented in GIS environment to have spatial distribution of crop output. In addition to the plant growth routine, the model includes components for weather simulation, hydrology and crop management. It operates on a daily time-step. Among factors simulated by EPIC are evapotranspiration (based on Penmen Monetih model), soil temperature, crop potential growth, constraints and yields. EPIC uses a ssingle model for simulating all crops, although, of course, each crop has unique values for model parameters. For this study, CO2 concentration held at current ambient level (350ppm) has been used. Crop yields are estimated by multiplying the aboveground biomass at maturity (determined by accumulation of heat units of specific field harvest data) by a harvest index for the particular crop.

Preparation of spatial-EPIC Dataset

Model requires data on soil properties, which has been assumed same for all the area as Spatial-EPIC is under progress and data required for these parameters are under development. The detailed profiles of relevant production characteristics (for example, fertilization, planting, harvesting, irrigation and heat units for maturity) were complied for the crops simulated across the region.

The weather variables necessary for driving the EPIC, a main impetus of this chapter, are daily values of precipitation, minimum/maximum are temperature, solar radiation and humidity.

DATA GENERATION: INTERPOLATION AND INTERPRETATIO

Most simulation model assumes homogeneous conditions over the space they are representing. In general, these conditions do not exceed a few square kilometers in extent. To input weather data across India, for example, we were faced with the problem of extending required model input over several hundred square kilometers between weather stations.

Several interpolation procedures are available, ranging from simple linear interpolation techniques and triangulated networks, to more sophisticated distance weighing or kriging techniques. It must be remembered that the only information used by any interpolation procedure is the location of known values. For this study, the climatic information used to make interpolation is based on inverse distance weighted method of more than 200 stations in India from which Bihar has been delineated for the purpose. While interpolation of a value at reach cell in the study area using 200 meteorological stations is technically easy, but some important questions can be raised: are the stations representative of the areas around them? How large or small an area do they represent? How does the spatial feature in question change over space? Is it continuos or discontinuous and abrupt?

We used same climate dataset for the different maps resolution across study area. Depending on its cell size, the level of detail becomes available at different resolution. It is thus important to realize the size of the dataset used to create GIS maps (as model input) and is important for model application from its accuracy and better performance view point. On the other hand a total of one or two stations falling in each administrative boundary may not be enough to give great confidence in the resulting interpolated map.

Results

Climatic data at 10 km spatial aggregation needs to be evaluated to identify its applicability in crop modeling; result of those are presented here. Observed yields were compared against simulated yields for two different grid cells of these there crops in the study area. As a mapping tools, GIS has been used for the production of descriptive maps using simulation model outputs that portray the spatial distribution of model prediction and the spatial juxtaposition of specified subsets of the data.

The geographical locations of the modeled process of different crops are simulated actual yield but there is a lack on observed data to tie the results to a real world location. An example of comparing spatially distributed simulated crop yield with observed aggregated crop yield at district levels. The above results suggest that the aggregation of climate data alone showed pertinent agreement between observed and simulated yield, as the other parameters except climate remain unchanged. Observed data available for validation are average of district administrative boundary, whereas simulated output varies cell by cell as of their resolution. To check its validity more explicitly we averaged the simulated yield at district level as zonal mean productivity and then compared with observed average. Sample of linear regression between observed vs simulated crop yield of rice, wheat and maize in the year 86, 89 and 91 respectively.

The outputs obtained on climate data gave quite satisfactory result and its is supposed that after introduction soil, topography and other input parameters spatially the model will yield a better result.

Therefore, result reinforces the conclusion that in the range of spatial output examined here, agreement between observed maize, wheat and reice with simulated Spatial-EPIC model is sensitive and it depends a lot on the climate data.

Analytic operation allow GIS to generate new data correlation and prediction of future conditions. Research is still being conducted and preliminary results from this study are just becoming available.

However, it is encouraging to have found a methodology that Provides a plant physiologist, a modeler and a GIS users a common ground to discuss simulation results and further potential research directions.

Simulated crop yield maps generated can be used to better communicate model predictions. Hence using this methodology a region/nation can be modeled for crop productivity, which help researchers and decision-makers understand the status and extent of climate effects on global processes such as rice, wheat and maize production.

As of now, the most difficult problem countered in applying Spatial-EPIC areas in the developing world is the apparent lack of quality environmental data availability, detail validity, comprehensiveness, and scale must be carefully considered before this methodology can be successfully implemented.

ELIMINATING THE EFFECT OF FOOD PRODUCTION ON THE CLIMATE

Agriculture accounts for between 10-12% of total human greenhouse gas emissions, with 60% of methane (CH_4) emissions, and 50% of all human nitrous oxide (N_2O) emissions. While food production's share of greenhouse gasses is small overall, its share of both nitrous oxide and methane is large. These gasses have a far higher effect than carbon dioxide which is by far the most prevalent gas.

Methane is about 21 times more effective at trapping heat than CO_2 (EPA, 2007), and nitrous oxide is far more than that, being 310 times more effective than CO_2 (EPA, 2006). From this, we can see that the effect of food production, while by no means the single source of Global Warming Potential (GWP), has a major effect. In addition to adapting our agricultural system to continue to provide us with secure sources of food in a warmer world, we must also adjust our agricultural practices and implement technology to reduce the affect of food production on the problem of Climate Change itself. Mitigation practices fall under several categories; reducing emissions, enhancing removals, and avoiding (or displacing) emissions. The first category involves changing practices to put less GHG's into the atmosphere. Changing feed practices for livestock and altering the method of nitrogen application to crops are two examples of practices that would reduce emissions of methane and nitrous oxide respectively.

Enhancing removals deals with improving the carbon sequestering potential of food production. According to the IPCC's Third Assesment report (2001), agricultural ecosystems have a high carbon storage potential, mostly in organic matter within the soil. Processes that increase photosynthesis and slow re-release of carbon via plant respiration or erosion will affect this result. Such stores of carbon are referred to as "carbon sinks". While sequestering other gasses is possible on a more limited level, the primary sequestering potential for CO_2- is sometimes at a cost of more emissions of other GHG's; such as methane or nitrous oxide. Once again, the methods and technologies used are expected to vary widely by local climate and geographical conditions.

The final method, avoiding or displacing emissions, involves using agricultural bi-products and crops as a source of energy and replacing fossil from fossil fuels or other sources. These "biofeedback" energy sources still release GHG's, however they gain benefit from the fact that the carbon used was recently removed from that atmosphere (an active part of the carbon cycle), and not from fossil fuels, where it was stored underground for centuries. With this sort of energy source, the only 'new' carbon introduced to the cycle is that used to produce and process the energy source. Other options, such as using methane from animal waste products to power generators, takes methane that would otherwise be emitted directly to the atmosphere and uses it to generate electricity.

Reducing emissions from croplands, especially of methane and nitrous oxide, can often be effected by altering the way the crops are managed. Crop management can also effect how much carbon is taken in by the plants. Adding nutrients to the soil, for example, can improve the carbon gaining potential of the plants.

But too much fertilizer can be a bad thing, increasing the emissions of N_2O. Precise regulation of fertilizer application can greatly reduce excess N_2O emissions; as can strict control of leaching. Periodically draining rice paddies could help control high methane emissions. Unfortunately, this could cause a spike in N_2O production from those same fields. Many of these practices have tradeoffs; either in terms of increased emission of another greenhouse gas, or damage to ecosystems in other ways. Loss of forests (a better carbon sink than croplands), due to increased agricultural production would clearly be a bad thing; while increased efficiency of production on current cropland (with possible spike in emissions) could slow the inevitable growth of agricultural lands, and prevent emission-high deforestation. Implementation of all of these practices must consider these factors and many more. A long view is needed; many of these practices change over time.

The majority of research on agricultural mitigation has been in the field of soil carbon sequestration. The IPCC Second Assesment Report (1996), estimated that 400-800 million metric tons of carbon could be sequestered in global agricultural soils per year, with saturation (and the end of sequestering) occurring after 50 to 100 years. Further, it was estimated that by using 10-15% of agricultural land to grow "energy crops", *i.e.* crops for ethanol or biodeisel, could offset 300-1300 million metric tons of carbon (or 1100-4800 metric tons of CO_2) per year from fossil fuels, by capturing previously emitted carbon and burning it again. Simply turning crop residue (*i.e.* stalks of plants, straw, etc.) into biogas, or direct burning it, could contribute 100-200 million tons of carbon (400-700 Mt CO_2), without disrupting food production .

ENSURING GLOBAL FOOD SECURITY

There are many ways to affect climate change and secure our future food supply. However, the future is very uncertain.

The IPCC makes recommendations, sometimes contradictory, but in the end, much of this is new ground. The IPCC recommendation is to use agriculture for producing biogas, but widespread production of ethanol in developed countries, along with incentives by national governments to produce ethanol, has been discovered to be a major contributor to the large rise in food prices in 2007 and 2008 by World Bank Economists and others. For reasons probably unrelated to Climate Change, China is undergoing severe agricultural decline.

They have lost one fifth of their agricultural land since 1949 due to soil erosion and economic development (CIA, 2008). China's imports of grain are growing as their population grows and environmental problems impact agriculture, and they are projected to have a grain production deficit on the order 40 million tons per annum. With China and other industrialized nations with grain deficits hunting for scarce food exports, the rise in prices is hitting poorer nations particularly hard. Countries with grain surpluses have been halting exports to guarantee food to their own populaces, as surplus food becomes scarcer. African nations are among the hardest hit; Mauritania has seen a 67% increase in the cost of wheat, a 117% increase in the cost of cooking oil, and a 25% increase in the cost of rice (the three key ingredients for food in the nation).

Mauritania currently has a roughly 45 day supply of these commodities; and producing nations are being stingy with imports. Wealthy importers such as Malaysia and Singapore are hording food and banking against even higher prices. People are starting to leave their homes in search of work and food. Even the United States is feeling the crunch; the price of wheat flour has doubled in just a few months, probably linked to a 24 million acre drop in wheat planting since 1981; some of which is due to the production of ethanol, and corn and soy for export.

With all of these shortages happening now, we can only wonder what climate change is going to do to an already dire situation. It may be true that climate change is already be affecting production in a limited way, but by all accounts, it will get far worse despite our best efforts. What will the world look like then?Food supplies and poverty levels have been significant global issues for centuries. The concept of achieving total food security, defined as "access by all people at all times to enough food for an active, healthy life," is one of the most challenging goals for humanity. However, with the recent onset of global climate change, providing enough food to sustain human populations is becoming increasingly difficult.

Most models show that global warming will reduce agricultural output, leading to shortages in many developing countries. With less supply come higher costs, and in turn less availability to already impoverished regions. Studies show that the number of hungry people in developing countries will increase by 1% for every 2-2.5% increase in prices. In 1980, there were an estimated 500 million people at risk of hunger. In 2060 this number is projected to be around 640 million, ignoring the effects of global change. The increase in cost due to lower yields is expected to increase this statistic by anywhere from 60 to 350 million people.

For researchers modeling the impact of climate change on poverty levels, agriculture has been the most significant variable. This is because many of the staple crops used to feed human populations are naturally more vulnerable to climate shifts. Studies have predicted agriculture's role in reducing poverty will decrease, leading to around 600 million additional people with inadequate food supplies. There is some debate as to whether small or large-scale farming would adapt better to global warming. While small farming are generally thought to be good adapters, they would not fare as well long term in a shifting environment, as they cannot afford to take risks.

In creating a viable and enduring agricultural system for the 21st century and beyond, great care must be taken to craft agricultural systems that take into account the dual problems of increasing population and food demand, and climate change. The problems are multivariable, interrelated and profound; and there is no clear answer facing us at this time. But at the same time, these problems are not going away, and much work and research is still needed to see us through.

Though most research considers a growing population a given of these agricultural problems, we feel that governments singularly and as a whole should begin to consider widespread population control measures. Population is growing, especially in developing countries, some of which are the hardest hit by the growing food crisis, and curtailing growth would reduce much of the strain to continue expanding agricultural production in the face of climate change and adverse effects such as deforestation. We believe that by implementing moderately strict population control along with these solution and mitigation practices, we stand a good chance of being able to feed our people for the next century. Humans have always faced the problem of feeding themselves, but never has the scope of the situation been so wide, and the solutions so grand and inspiring.

CONSERVATION VS FOOD PRODUCTION: A FALSE DICHOTOMY

It is tempting, given the question's use of the term "conserving ecosystems," to focus this answer on the 14% of the terrestrial system that is protected. Such a response, however, would miss the elephant in the room. To conserve ecosystems in a changing climate, we must instead focus on how they are used and managed in the remaining 86% of land cover. Food production currently dominates that use, with 40 percent of the Earth's land cover managed for crop production and pasture. To date, the focus on managing ecosystems for food production has resulted in significant benefits for humans. The Millennium Ecosystem Assessment found that globally human well-being has improved over the past 50 years, as measured by the UN human development index. We live longer, are better nourished, literacy rates have risen along with educational attainment, and per capita GDP is rising.

Yet, as human well-being has improved, the health of ecosystems has declined. Humans have degraded the majority of ecosystem services""capture fisheries, freshwater, pollination and many more""while enhancing a few food services. Food production is a major force behind every direct driver of ecosystem degradation identified by the Millennium Ecosystem Assessment in 2005.

Food production as a major driver of ecosystem degradation Habitat conversion: Approximately 43% of tropical and subtropical forests and 45% of temperate forests have been converted to croplands.

Over exploitation: 70% of global freshwater use is by agriculture, reducing its availability for other uses.

Invasive species: The introduction of aquatic alien fish species has led to the extinction of native species in many parts of the world.

Pollution: Only a fraction of nitrogen applied as a fertilizer is typically used by plants, the rest ends up in inland waters and coastal systems, creating eutrophication and dead zones.

Climate Change: Agriculture directly contributed to around 14% of global greenhouse gas emissions in 2005 and drives additional emissions through its role in deforestation.

Minqin Oasis in Western China, for example, historically served as a natural barrier against the dryness of the Tengger and Badain Jaran deserts. In the 1950s Chairman Mao implemented a national plan to boost food production, entailing cultivation, deforestation, irrigation, and reclamation. The long-term consequences on other ecosystem services such as erosion control and freshwater were devastating. As a result, the Minqin oasis has been swallowed by deserts, leading to abandonment of once productive land. In the U.S. midwest region, the production of crops, livestock, and biofuel has degraded coastal ecosystems. In the Gulf of Mexico, nutrient runoff from farmland has created a dead zone approximately the size of New Jersey which has degraded fisheries and is expected to worsen with climate change.

Can this paradox of improving human well-being by increasing food production at the expense of other ecosystem services that humans depend on continue? Scientists worry that time lags may exist between the degradation of ecosystems and the resulting effects on human well-being. And that these may lead to tipping points past which damage to ecosystems is irreversible. The Millennium Ecosystem Assessment identified several examples of irreversible ecosystem change, such as the sudden collapse of chronically over-exploited North Atlantic cod stocks in the late 1980s. As climate change impacts exacerbate food production stresses on ecosystems, it is conceivable that such collapses become commonplace, with enormous implications for food security, especially in the developing world where two billion rural poor depend on ecosystems for sustenance and livelihoods. For example, the Amazon provides critical water and climate regulation services that the region's agricultural sector depends upon for its survival.

Yet scientists fear that ongoing deforestation combined with climate change could bring the Brazilian Amazon to a tipping point beyond which it experiences widespread dieback and transitions into savanna-like vegetation. The reductions in rainfall would devastate efforts to raise crops and cattle in the region.

Further upping the food production/ecosystems health challenge facing policymakers, population growth and rising per capita incomes are expected to double the world's demand for food in the next 40 years, according to UN food and agriculture chief, Jacques Diouf.

Avoiding Ecosystem Services Tipping Points

To conserve ecosystems in the face of climate change, confronting the elephant-the way we meet our needs for food-must be a prime focus. And our approach to conserving ecosystems must ensure that they are not pushed beyond dangerous tipping points. Future approaches to conserving ecosystems must tackle the three interlinked challenges of climate change, ecosystem services degradation and rising demand for food.

Climate change: Climate change affects the quantity and quality of food and other ecosystem services. Moderate warming is expected to benefit agriculture in mid to high latitudes, but to decrease yield of cereals in low latitudes (medium confidence). The loss of ecosystem services is likely to be a major source of surprises from climate change. While ecosystems have adapted before in response to shifts in climate, their resilience this time is undermined by existing and growing human pressures. Future food production must therefore reduce its pressure on ecosystems and plan for the unexpected, such as changes to freshwater, natural pest regulation and pollination services.

Ecosystem service degradation: The majority of ecosystem services globally have been degraded, including those that underpin food production. Crops, livestock and aquaculture are among the few ecosystem services that have been enhanced. The degradation of ecosystems services is expected to grow significantly worse in the first half of the 21st century, with climate change possibly becoming the dominant driver by the end of the century.

Doubling of Food Demand

The focus on increasing food production needs to continue, if we are to sustain previous gains in human wellbeing, feed a growing population and lift millions out of poverty. But increasing food production at the expense of other ecosystem services and climate change cannot be sustained indefinitely. Approximately 13 million ha of forest, for example, are lost annually, primarily as a result of the expansion of land used for food and fuel. Deforestation is a significant cause of climate change and results in the loss of ecosystems services that are critical inputs to agriculture, including erosion control, climate regulation, pollination and water regulation. Yet, projections estimate conversions of an additional 120-240 million ha by 2030, much of it in Latin America and Sub-Saharan Africa and much of it for food production.

Addressing the intertwined challenge of meeting rising demand for food and conserving ecosystems in a changing climate will not be easy. It will require policy makers at all levels of society adopting different approaches to conserving and managing ecosystems that factor in climate change and other drivers of ecosystem change. Two such approaches are described below: 1. integrate climate change and ecosystem service risks in decisions that affect ecosystems and 2. reconcile food production and conservation goals.

Integrate climate change and ecosystem service risks in decisions. Future ecosystem use decisions, particularly those that involve food production, should assess the risks associated with their dependence and impact on ecosystem services and how these risks will be altered by climate change. A rapid assessment tool for this purpose, based on the Millennium Ecosystem Assessment conceptual framework and the World Resources Institute's subsequent work.

This comprehensive tool identifies the direct drivers of ecosystem change, including climate change, in a particular setting, and assesses both risks and opportunities stemming from a decision's dependence and impact on ecosystem services. It can be applied by policy makers working in different sectors and at multiple scales, from field to forest or watershed from to national to regional and local levels.

Reconcile food production and conservation goals. Future approaches to food production must first and foremost recognize its interdependence on a basket of ecosystem services and develop strategies to sustain them in the face of climate change. To achieve this, food security experts need to work alongside agriculturists and biologists to maximize food production while minimizing further ecosystem and biodiversity loss. Three key strategies can help meet this goal:

Restore Degraded Lands

There is an urgent need to halt the expansion of food production into natural ecosystems in order to conserve ecosystem services and biodiversity. Globally, over one billion hectares of land is believed to have restoration potential for increased human use-a global combined area greater than China. More work is needed to determine which of this is suitable for food production, taking into account water availability, climate change and social factors.

Restoring even a small part of this for food production would help reduce pressure on natural ecosystems. In Indonesia, for example, the World Resources Institute is seeking to develop a scalable model for diverting new oil palm plantations that would otherwise replace virgin forests onto degraded land. This ensures that oil palm plantations can keep expanding to meet demand""generating local revenues and jobs while reducing deforestation. Similar opportunities exist to divert the expansion of cattle ranches in the Amazon region to degraded lands.

Increase Productivity on Existing Farmland

While intensification doesn't immediately come to mind when thinking about conservation, it is nevertheless a key strategy to reduce stress on natural ecosystems. The challenge is to find ways to produce more food without unwanted ecosystem trade-offs. As noted by Shivaji Pandey, Director of FAO's Plant Production and Protection Division at a 2009 Congress on Conservation Agriculture in New Delhi.

"In the name of intensification in many places around the world, farmers over-ploughed, over-fertilized, over-irrigated, over-applied pesticides. In doing so we also affected all aspects of the soil, water, land, biodiversity and the services provided by an intact ecosystem. That began to bring yield growth rates down." Food production needs to be redesigned to be both more productive and synergistic with other ecosystem services. Public subsidies and tax incentives, for example, should be used to compensate farmers for managing their lands in ways that sustain beneficial ecosystem services to society. To contend with climate change, farmers need location-specific information on ecosystem services and changing climate conditions. This should be combined with research on how improved practices can address climate constraints on crops. For example, rainfall may remain plentiful in parts of Africa but be concentrated in shorter time periods. Ways to minimize run-off and soil erosion and store water to extend growing seasons will thus be important.

Manage Demand for Food

Opportunities for managing demand for food in both developed and developing countries include changing diets, reducing food waste and advancing programmes that encourage sustainable food production. The world's livestock sector is growing at an unprecedented rate driven by rapid growth in demand for animal protein as a result of population growth, rising incomes and urbanization. Animal protein requires more land, water, and energy to produce than plant protein.

Education programmes can help consumers make informed choices on the environmental and health benefits of food choices. There is also a need to spur and scale up demand for sustainably produced food. This can include promoting the use of certification and labeling programmes for sustainably produced agriculture commodities. If, by 2050, the world celebrates success in providing food security and in navigating ecological tipping points in the face of climate change, it will be because of the ingenuity of farmers and conservationists, agricultural experts and ecologists in finding ways of learning and acting together across scales and disciplines. In addition, political leaders will have supported much larger investments in research and training, and in monitoring the interaction of agriculture with ecosystem services and how the supply of services may change and be sustained in the face of climate change. The good news is that examples of "tomorrow's approach" are already beginning to emerge. The challenge now is to scale them up in a changing climate.

CORRELATION OF SPECTRAL REFLECTANCE

Multispectral data from aircraft or satellite have been widely used in agriculture as well as in other industrial sectors. To connect such data with vegetation cover for agricultural applications, generally a detailed and through ground-based should be made to characterize radiance and reflectance spectra and link it with growth and yield traits. This study was to remotely measure and monitor the seasonal variations in reflectance spectrum, in the range of 350-1100 nm, of rice (Oryza sativa L. cv. Tainung 67) canopy at the experimental field of Taiwan Agricultural Research Institute (TARI) during the cropping seasons in 1996-1997. The objectives were to examine relationship between growth traits and NDVI and to assess their potential use in estimating growth of rice crop. Attempts were also made to compare NDVI calculated from narrow-band ground-based spectral measurement and the simulated board-band multispectral satellite data. The physiological and morphological implications of the relationships were discussed.

Satellite or aircraft remote sensing can provide a timely spatial distribution information on crop performance and be incorporated into farm management scheme. Application such as precision farming activities, growth and production estimation, and land surface characteristics evaluation are made possible. Before these practical applications, vegetation spectral properties from remote sensing must be in connection with some physical characters of plant canopy so that plant performance can reasonably be assessed. Mathematical formulae are then established from the fitting curves of these paired variables. Leaf area index (LAI), percent vegetation cover, and unit area plant biomass are among physical characters commonly used as indicators of the status of crops.

Vegetation indices are mathematical transformations intended to estimate the spectral contribution of crop vegetation to multispectral observations. The transformations have the effect of normalizing measurements acquired from varied environmental conditions. The formulae are derived mostly from discrete green, red and near-infrared bands, especially the two ends of the so-called chlorophyll red-edge. By correlating vegetation indices with physical characters of plant canopy, changes of vegetation feature can be potentially assessed and predicted from values of vegetation indices during the growing season.

In this study, variations in reflectance spectrum were monitored and analyzed and the normalized difference vegetation index (NDVI) were calculated from the ground-based remotely sensed spectral measurements of rice canopy for the growing seasons of 1996-1997. An simplified vegetation model for estimating crop growth traits from NDVI is presented, and the simulated broad-band versions of NDVI were compared with the narrow-band counterparts.

MATERIALS AND METHODS

Experiments were conducted at TARI Experimental Farm (23o30'N, 120°42' E, elevation of 85 m) on a loam soil (nonacid, hyperthermic, Fluvaquentic Dystrochrept) during First and Second crops of 1996 and 1997. Cultivation and fertilization were in consistent with the existing cultural practices for rice adopted by the local region. Each season had 3 plots as replications and were furrow-irrigated. Herbicide and hand-weeding were practiced to avoid weed interference. Plants were sampled from 2 wk after transplanting till harvest periodically. Leaf area was determined and LAI, defined as the area of green leaves per unit area of land, was computed at each sampling. Leaf dry weight (LDW) was measured after oven-dried at 80oC for 72 h. Means were used for regression analyses to examine the correlation between growth traits and NDVI.

Method of spectral measurements was reported in the previous papers. Spectral irradiance of the dense vegetation cover and incident solar radiation in the range of 350 to 1100 nm was measured using a LI-1800 high spectral resolution portable spectroadiometer (LI-COR inc., Nebraska, USA). The sensor head was pointed downward in a nadir-viewing and was placed horizontally 1.0 m above rice vegetation surface to scan the upward reflected radiation. Measurements were made at each plot using sunlight as illumination source on clear or near cloudless days between 11:00 to 12:00 local standard time. Date were averaged and mean values were used.

Radiance spectra of sunlight were taken before and after each replicate measurement and the average was used as the reflectance standard. Reflectance of individual wavelength was calculated by dividing the vegetation radiance measurements with the corresponding incident solar radiation measurement with the corresponding incident solar radiation measurements. Reflectance spectra measured during the growth were grouped for each crop and the standard errors of individual wavelengths were calculated as representative of the upper and lower boundaries. The NDVI was defined as (NIR-RED)/(NIR+RED). The reflectance of the wavelengths at RED (674 nm) and NIR (754 nm) measured from LI-1800 were selected and used for calculations. Regression analysis were performed to generate fitting-curves in order to monitoring seasonal changes of NDVI as well as growth traits.

From narrow-band reflectance spectra, broad-band RED and NIR reflectance were computed. The reflectance simulated for RED and NIR for HRV sensors of French SPOT programme were derived from the 2-nm-wide narrow-band LI-1800 spectra in bandpasses of 610-680 nm and 790-890 nm. For MSS sensor system of Landsat series, bandpasses in the range of 600-700 nm and 700-800 nm were computed to simulated Band 2 and Band 3, respectively. The averaged values were presented for RED and NIR bands of the broad-band sensors. Linear regression was performed to compare the difference in NDVI between values calculated from single reflectance and band reflectance.

Results and Discussion

For a typical reflectance spectrum, wavebands in the visible region were mostly absorbed by chlorphylls while wavelengths in NIR were reflected significantly because of the absence of absorption. It was shown similar in seasoanl pattern of reflectance spectrum but was shifted by weather and soil impact. Variations were found much greater in the early growth and maturing stages than during the long middle period of the growing season. The greater standard errors graphed in the 1st crop of 1997 inferring a stronger climate oscillation occuring in the growing period. These 'environmental' and 'growth' effects influencing the time-seqential performance of canopy reflectance spectra were also found in many field crops.

The NDVI reached a plateau when canopy cover was complete. During late season, decline of NDVI was responsive to the onset of chlorosis due to crop maturation. For a précised estimation of canopy status, relationship between growth traits and NDVI should be carefully classified. Because the relationship are adversely influenced by senescence, it would be better not to assess growth traits in the early and the late stages of plant development. As the sustainability of full vegetative cover is an important factor in determining the final yield, the spectral assessment of the development stages is crucial.

To examine whether satellite broad-band data can be used to substitute for ground-based narrow-band measurements, a linear regression analysis was performed. The broad-band versions of NDVI agree very well (R^2=0.982 for SPOT and R^2=0.985 for LANDSAT) with the narrow-band counterparts. As these spectral transformations are highly correlated, estimation of growth traits based on NDVI can be fairly accurate for both narrow-band data and spectral inputs collected from satellite. Thus, the reflectance band of the HRV and MSS sensors could be used as predictors of canopy characters. With many more interacting variables, spectral data measured from satellite remote sensing are considered more complicated than those from the ground platform.

NATURAL ENVIRONMENT AND FOODS

Among the nine reported species of Carnobacterium, only two species, C. divergens and C. maltaromaticum, are frequently isolated from various sources. Four species, C. alterfunditum, C. funditum, C. gallinarum and C. mobile, have been isolated a few times from at most three sources. Three species, C. inhibens, C. pleistocenium and C. viridans, have only been isolated from one source, but it should be noted that Carnobacterium spp. related to, for example, C. inhibens have been described. Carnobacterium spp. appear to have both the temperate and polar aquatic environments as habitats including live fish, marine sponges, Antarctic lakes, Arctic and Antarctic sea water as well as the deep sea, aquous alkaline tufa columns from Greenland, and freshwater habitats from the temperate clima zone, including a Sphagnum pond and rivers in the northwest region of Spain.

Although C. maltaromaticum and/or C. divergens have been isolated from tropical fish products, including smoked surubim, a Brazilian tropical freshwater fish, and from vacuum-packed tuna caught in the Indian Ocean and processed in Sri Lanka, it cannot, at least in the latter case, be excluded that this is due to contamination associated with repackaging in Denmark.

The presence of carnobacteria has also been demonstrated in the terrestrial environment, including a field treated with whey, Canadian winter soil, permafrost ice, a compost pile, a collapsed horse manure pile, the larval midgut of a moth species, and other sources.

It appears that the temperate/polar aquatic and terrestrial environments are both natural habitats. Carnobacterium divergens and C. maltaromaticum possess traits that may play a role in their survival in these surroundings. One study has reported that a Carnobacterium sp. soil isolate related to C. maltaromaticum survived 48 serial freeze-thaw cycles better than Escherichia coli and an Enterococcus sp. soil isolate, but worse than a few other soil isolates, including an Acinetobacter sp. Indeed, these organisms may survive freezing for considerable periods of time, as witnessed by the isolation of a Carnobacterium sp. preserved in a permafrost ice wedge for 25 000 years.

Additional data on the abilities of carnobacteria to grow at low temperatures and to survive frozen storage is available for food products. Also, a cold-active b-galactosidase from C. maltaromaticum and a cold-adapted alanine dehydrogenase from a Carnobacterium sp. related to C. alterfunditum have been reported. Some carnobacterial isolates originate from natural high-pressure habitats, and additional data on resistance to pressure have been reported for high-pressure-processed foods. It has not been described how other physical/chemical parameters such as salt content, atmosphere and pH affect survival and growth of these organisms in the natural environment.

With respect to energy-yielding substrates, one species, C. maltaromaticum, expresses chitinase activity. This may facilitate adherence to and survival on zooplankton as suggested for enterococci, and also explain the isolation of this species from the midgut of the larval stage of a species of moth. Interestingly, the arginine deiminase pathway is expressed by the most frequently encountered species, C. divergens, C. gallinarum, C. maltaromaticum and C. mobile but not by C. inhibens or C. viridans. Arginine is not a substrate that results in growth of C. alterfunditum and C. funditum, and information is not available for C. pleistocenium. This amino acid may represent an additional energy source for growth and survival when carbohydrates are scarce, and it may offer protection against acid stress, as described for Enterococcus faecalis and related species. Finally, carnobacteria are able to catabolize a range of carbohydrates, although there are considerable interspecies and intraspecies heterogeneities.

Examples of sources of such carbohydrates include animals (*e.g.* chitin and, to some extent, lactose, although the targets of the lactose hydrolytic activity shown by carnobacteria may instead be byproducts of plant sugar polymers and saccharides adsorbed to humic acid substances in the soil), plants (*e.g.* salicin and sucrose), fungi (*e.g.* chitin and trehalose), prokaryotes (N-acetylglucosamine), and living organisms in general (ribose). The importance of these abilities for growth and survival in the environment deserves further study. The genome sizes of C. alterfunditum, C. divergens and C. pleistocenium strains and a Carnobacterium sp. AT7 deep sea isolate have been estimated to be 2.9, 3.2, 3.2 and 2.4Mb, respectively, and for the AT7 isolate, the data are reported in a database described by Liolios (2006). These sizes are relatively large in comparison to many other LAB, suggesting that carnobacterial genomes may encode a range of genes that makes them well adapted to deal with environmental challenges. Even if carnobacteria are well adapted to temperate or polar aqueous environments, this does not always provide improved ability to survive as compared to other bacteria. Thus, the actual abilities of strains of C. divergens and C. maltaromaticum isolated from a Sphagnum pond to survive in water from this source were not improved as compared to related gram-positive bacteria originating from other sources. At present, therefore, the precise mechanisms by which Carnobacterium spp. persist in the natural environment and their underlying genetics are not known. Finally, it should be noted that the environments from which carnobacteria can be isolated are not always as extreme as they appear at first sight. Thus, Carnobacterium spp. isolated from Lake Vanda, Antarctica were found at a depth of 61 m, where the temperature was c. 15-20 1C.

DAIRY, FISH AND MEAT PRODUCTS

High concentrations of bacteria (4106-107 CFUg_1) in food are typically required before their activity is sufficient to influence the sensory properties of a product. For this to happen, occurrence and growth kinetics are the key parameters, and databases, including ComBase, are available to determine the effect of food storage conditions and product characteristics on growth of specific bacteria. However, information on carnobacteria has not yet been included in such databases.

Carnobacterium maltaromaticum and another Carnobacterium sp. have been isolated from milk, and they, in addition to C. divergens, have been detected in soft cheeses, including mold-ripened brie, mozzarella, Camembert and other types. The high concentration of C. divergens in the curd of mozzarella, exposed to 10-37 1C during processing, is in agreement with the maximum growth temperature (40 1C) of this species.

The occurrence of Carnobacterium in dairy products (and other foods) is most likely underreported. This is due to the common use of acetate containing media, particularly MRS agar or Rogosa agar, for enumeration of LAB. Growth of Carnobacterium is inhibited by acetate, and both MRS and Rogosa agar media significantly underestimate concentrations in food.

Carnobacterium divergens and C. maltaromaticum are present in seafood and are able to grow to high concentrations in different fresh and lightly preserved products.

Studies of naturally contaminated products suggest which storage conditions and product characteristics select for carnobacteria as compared to the other bacteria present in seafood. For chilled fresh seafood, we have found no reports where C. divergens and C. maltaromaticum dominated the microbial communities in aerobically stored products, but this has been reported for modified atmosphere-packed (MAP) coalfish, cod, pollack, rainbow trout, salmon, shrimp, swordfish and tuna.

After frozen storage (_20 to _30 1C for 5-8 weeks), C. divergens and C. maltaromaticum seem to be particularly prominent in chilled MAP fish, as has been reported for cod, garfish, salmon, and tuna. In addition to freezing, these carnobacteria are relatively resistant to high-pressure processing and are found in high concentrations in vacuum-packed and chilled squid mantle and cold-smoked salmon previously treated with 200-400MPa for 15-20 min. In vacuum-packed coldsmoked or sugar-salted ('gravad') seafood with 3-7% NaCl in the water phase and a pH of 5.8-6.5, high concentrations of C. divergens and C. maltaromaticum are particularly common, as reported for halibut, rainbow trout, salmon, surubim, and tuna.

High concentrations of C. maltaromaticum are also reported in salted lumpfish roe, and it has been isolated from frozen, smoked mussels. Finally, for cooked seafood, high concentrations of C. maltaromaticum have been detected in MAP shrimp after storage at 2-8 1C. Both C. divergens and C. mobile were isolated from cooked and brined MAP shrimps (with NaCl, benzoic acid, citric acid and sorbic acid) after storage at 2-8 1C. Clearly, carnobacteria are common in chilled fresh and lightly preserved seafood, but at higher storage temperatures (15-25 1C), other species, including Enterococcus spp., more frequently dominate the spoilage microbial community of seafood.

Carnobacterium divergens and C. maltaromaticum are able to grow in meat products at temperatures as low as 2 to _1.5 1C, and they are frequently predominant members (up to 50% C. divergens and up to 26% C. maltaromaticum of the gram-positive or LAB isolates obtained) of the microbial community of raw meat (beef, pork, lamb, and poultry).

The two species are found irrespective of whether products have been stored aerobically, vacuum packaged, or subjected to modified atmospheres, including gas compositions of CO_2/N_2 (%) ranging from 10: 90 to 80: 20. One study demonstrated the presence of C. divergens in beef stored in air but not in MAP beef with increased concentrations of O_2 (20-40%) in addition to CO_2 (40%). The growth of C. funditum is impaired by oxygen, and it will be of interest to study further whether addition of O_2 to the gas composition of modified atmospheres consistently inhibits growth of carnobacteria, as has been observed for some other bacteria.

Further studies are needed in order to determine whether differences in the presence of carnobacteria in meat are due to variations in storage conditions or variations in contamination levels at the processing plants. The source of carnobacteria in meat products is most probably the processing plant, as these organisms have not been isolated from the gastrointestinal system or skin of chicken, cattle, pigs or sheep, except for one unpublished study. Thus, the source of contamination of broiler carcasses by C. divergens and C. maltaromaticum was shown to be the air in the processing plant and not incoming broiler chickens. In fact, two studies confirmed that C. divergens and C. maltaromaticum present in the raw material were eliminated or reduced in number by cooking of a processed meat product, ham.

Carnobacterium divergens and C. maltaromaticum have, however, been detected in a variety of processed meat products, including the cured pork product bacon, a Danish processed pork product ('rullep_lse'), various Spanish processed meat products, cooked poultry meat, pressure-treated (408-888MPa) chicken, and irradiated pork and chicken. On a few occasions, C. maltaromaticum has even been isolated from fermented sausages; as noted by Hammes & Hertel (2003), this contrasts with the usual nonaciduric habitats of carnobacteria. Other Carnobacterium spp. isolated from processed meat products include C. gallinarium (irradiated chicken), C. mobile (cooked turkey) and C. viridans (vacuum-packed Bologna sausage). Although these organisms are frequently isolated from processed meat products, it has been suggested that they are rarely present in high numbers, and that terminal spoilage of cooked, cured meats is primarily caused by aciduric LAB.

Finally, a Carnobacterium sp. has also been isolated from egg contents, and it was shown that this isolate had a similar ability to penetrate the eggshell and contaminate the whole egg as various other gram-positive and gram-negative bacteria.

Transmission routes for carnobacteria from the natural environment into food-manufacturing plants and further on to dairy, fish and meat products are not known to any extent. Knowledge on this topic is essential to evaluate if species and intraspecific clusters that differ in spoilage ability also have different colonization abilities. Such knowledge would also illuminate the feasibility of using the production of metabolites, especially tyramine, by C. divergens and C. maltaromaticum as an index of microbial spoilage of specific fish and meat products.

MANAGEMENT FOR ENVIRONMENTAL PROTECTION

The irrigated agriculture is a major contributor to the salinity of many surface-and groundwaters. The agricultural community has a responsibility to protect the quality of these waters. It must also maintain a viable, permanent irrigated agriculture. Irrigated agriculture cannot be sustained without adequate leaching and drainage to prevent excessive salinization of the soil, yet these processes are the very ones that contribute to the salt loading of surface and groundwaters. But surface and groundwater salinity could be reduced if salt loading contributions from the irrigation processes were minimized or eliminated. The protection of water resources against excessive salination, while sustaining agricultural production through irrigation, requires the implementation of comprehensive land and water use policies that incorporate the natural processes involved in the soil-plant-water and associated geohydrological systems.

Strategies to consider in coping with increasing salinity in receiving water systems resulting from irrigation include:

- eliminating irrigation of certain polluting lands;
- intercepting point sources of drainage return flow and diverting them to other uses;
- reducing the amount of water lost in seepage and deep percolation;
- isolating saline drainage water from good quality water supplies.

MINIMIZING DEEP PERCOLATION AND INTERCEPTING DRAINAGE

The minimizing leaching always reduces the salt discharged from the rootzone. Additionally, deeply percolating water often displaces saline groundwater of higher salinity or dissolves additional salt from the subsoil. Reducing deep percolation will generally reduce the salt load returned to the river as well as reduce water loss. The "minimized leaching" concept of irrigation which reduces deep percolation should be adopted and implemented to reduce salinization of water resources associated with irrigation projects, especially in projects underlain by salt-laden sediments. In addition, saline drainage water should be intercepted. Intercepted saline drainage water can be desalted and reused, disposed of by pond evaporation or by injection into some isolated deep aquifer, or it can be used as a water supply where use of saline water is appropriate. Desalination of agricultural drainage waters for improving water quality is not generally economically feasible even though it is to be implemented for the return flow of the Wellton-Mohawk irrigation project of Arizona, USA. The high costs of the pretreatment, maintenance, and power are deterrents. Only in extreme cases, or for political rather than technical reasons, is desalination advocated.

Drainage Water for Irrigation

While there is an excellent opportunity to reduce the salt load contributed by drainage water through better irrigation management, especially through reductions in seepage and deep percolation, there are practical constraints which limit such reductions. But the ultimate goal should be to maximize the utilization of the irrigation water supply in a single application with minimum drainage. To the extent that the drainage water still has value for transpirational use by a crop of higher salt tolerance, it should be used again for irrigation.

Drainage waters are often returned by diffuse flow or intentional direct discharge to the watercourse and automatically "reused." Dilution of return flows is often advocated for controlling water salinity. This concept has serious limitations when one considers its overall effect on the volume of usable water, and it should not be advocated as a general method of salinity control. The preferred strategy to control the salinity of water resources associated with irrigated lands is to intercept drainage waters before they are returned to the river and to use them for irrigation by substituting them for the low-salinity water normally used for irrigation at certain periods during the irrigation season of certain crops in the rotation. When the drainage water quality is such that its potential for reuse is exhausted then this drainage should be discharged to some appropriate outlet. This strategy will conserve water, sustain crop production and minimize the salt loading of rivers that occurs by irrigation return flow. It will also reduce the amount of water that needs to be diverted for irrigation.

The ultimate goal of irrigation management should be to minimize the amount of water extracted from a good-quality water supply and to maximize the utilization of the extracted portion during irrigation use, so that as much of it as possible is consumed in transpiration (hence producing biomass) and as little as possible is wasted and discharged as drainage. Towards this goal, to the extent that the drainage water from a field or project still has value for transpirational use by a crop of higher salt tolerance, it should be used again for irrigation before ultimate disposal.

An integrated strategy of management that will simultaneously facilitate the successful use of saline waters for irrigation, minimize the harmful off-site effects of drainage discharge on the pollution of water resources and maximize the beneficial use of the total water supply available in typical irrigated lands and projects. This strategy illustrates how the information and principles given in the preceding of these guidelines can be integrated towards the goals of sustaining irrigated agriculture and protecting soil and water resources.

To the extent practical, water diverted and applied for irrigation should be minimized using the principles and methods. Unavoidable excessive, usable resulting drainage water should be intercepted and isolated from good-quality water supplies and used within dedicated parts of the project as a substitute for part of the freshwater given to the crops. The "dual rotation, cyclic" management strategy of Rhoades (1984a, b, c) can be used to enhance the feasibility of reusing such saline drainage waters for irrigation. In this system, sensitive crops (such as lettuce, alfalfa, etc.) in the rotation are irrigated with "low salinity" water (usually the developed water supply of the irrigation project), and salt-tolerant crops (such as cotton, sugarbeets, wheat, etc.) are irrigated with saline drainage water or the shallow groundwater created by over-irrigation in the project. For the salt-tolerant crops, the switch to saline water is usually made after seedling establishment; preplant irrigations and initial irrigations being made with low-salinity irrigation water. The secondary drainage resulting from such re-use should also be isolated and used successively for crops (including halophytes and tolerant trees) of increasingly greater salt tolerance. The ultimate unusable drainage water should be disposed of to some appropriate outlet or treatment facility.

The feasibility of this "dual-rotation, cyclic" strategy is supported by the following: The maximum possible soil salinity in the rootzone resulting from continuous use of saline water does not occur when this water is used only for a fraction of the time. Alleviation of salt build-up resulting from irrigation of salt-tolerant crops with the saline water occurs later when a salt-sensitive crop (s) is irrigated with the low-salinity water supply, or during off season periods of high rainfall. Proper preplant irrigation and careful irrigation management undertaken during germination and seedling establishment are made using the low-salinity water supply to leach salts accumulated from saline irrigations out of the seed-area and from shallow soil depths.

Data obtained in modelling studies and in field experiments support the credibility and feasibility of this "cyclic" reuse strategy. Results of an experiment to test the feasibility of the cyclic, "dual-rotation" reuse strategy are reviewed to clarify and illustrate the concept and to demonstrate its credibility. The strategy was tested in a 20 ha field experiment on a commercial farm in the Imperial Valley of California.

Two cropping patterns were tested. One was a two-year, successive-crop rotation of wheat, sugarbeets and cantaloupe melons. In this rotation, Colorado River water (900 mg/l TDS) was used for the preplant and early vegetative growth stage irrigations of wheat and sugarbeets and for all irrigations of the melons. The remaining irrigations were with drainage water of 3500 mg/l TDS (Alamo River water). The other cropping pattern tested was a four-year block rotation consisting of two consecutive years of cotton (a salt-tolerant crop) followed by wheat (a crop of intermediate salt-tolerance) and then by two years of continuous alfalfa (a relatively salt-sensitive crop). Drainage water was used for the irrigation of cotton after seedling establishment; beginning with the wheat crop, only Colorado River water was used. From Watsuit calculations, it was hypothesized that the crops irrigated with the drainage water would yield fully when established with Colorado River water and from other calculations that sufficient desalination of the soil would occur when irrigating with Colorado River water to achieve a good plant stand and to keep the soil from becoming excessively saline over the long-run.

No significant losses in the yields of the wheat and sugarbeet crops occurred in either cycle of the successive crop rotation from substituting drainage water (even in the greater amount; 65-75 percent; treatment cA) for Colorado River water for the irrigation of these crops after seedling establishment. The mean yield of cantaloupe seed obtained in the cA plots was about 10 percent lower than the control, but the difference was not statistically significant. The yields of the fresh-market melons (numbers of cartons of cantaloupes obtained by commercial harvest operations) in 1985 was higher in the Ca and cA treatments than in the C treatment, but they were not significantly different. Hence, no significant yield loss was observed from growing cantaloupes using Colorado River for irrigation in the land previously salinized from the irrigation of wheat and sugarbeets using drainage water. In the block rotation, there was no loss in lint yield in the first cotton crop (1982) from use of saline drainage water for irrigation, even when it was used for all irrigations, including the preplant and seedling establishment periods (treatment A).

There was no significant loss in lint yield in the second cotton crop (1983) grown in the block rotation from use of drainage water for the irrigations given following seedling establishment which was accomplished using Colorado River (the recommended strategy treatment, cA). But there was a significant and substantial loss of lint yield, as expected, where the drainage water was used solely for irrigation (the "extreme-control" treatment, A).

This loss of yield was caused primarily by a loss of stand that occurred this second year due to excess salinity in the seedbed during the establishment period. No loss in yield of the wheat grain or alfalfa hay crops occurred in the block rotation associated with the previous use of drainage water to grow cotton on these lands when they were subsequently grown with use of Colorado River water for irrigation. The qualities of all of these crops were never inferior, and often were superior, when grown using the drainage water for irrigation or on the land where it had previously been used. These quality data are given elsewhere.

The average amounts of water applied to each crop and over the entire four-year period, the successive and block rotations, respectively. These data include all water applied, including that used for preplant irrigations and land preparation purposes. That substantial amounts of drainage water were substituted for Colorado River water in the irrigation of these crops without yield loss.

The estimated amounts of water consumed by the crops through evapotranspiration and lost as deep percolation, individual crop and by succession of crops for both rotations. It was assumed that consumptive use was the same in all treatments, since no substantial losses of yield resulted in any treatment. These data show that the saline drainage water was successfully used for irrigation without resorting to high leaching. Data on levels of soil salinity and sodicity in the seedbeds and rootzones are given in Rhoades *et al.* (1989b). Their levels were kept within acceptable limits for seedling establishment and the subsequent growth of the individual crops grown in the rotation when the recommended strategy was employed. These results along with the high crop yields and qualities obtained in this test under actual farming conditions support the credibility of the recommended cyclic, dual-rotation (crop and water) strategy to facilitate the use of saline waters for irrigation.

In this cyclic strategy, steady-state salinity conditions in the soil profile are never reached, since the irrigation water quality changes with crop type in the rotation and with time in the irrigation season. Consequently, a flexible cropping pattern which includes salt-sensitive crops can be achieved. The intermittent leaching which occurs using this strategy is more effective in leaching salts than is continuous leaching (*i.e.* imposing a leaching fraction with each irrigation) for the reasons given earlier. Another advantage of the strategy is that a facility for blending waters of different qualities is not required.

In order to plan and implement a successful practice involving the use of the cyclic, dual-rotation strategy for irrigating with saline waters, various other considerations must be addressed. The intention here is not to provide a step-by-step process that must be followed nor a rigid set of criteria to address these considerations, since most management decisions are subjective and case specific, but to discuss some of the factors that should be considered and to provide some rough guidelines for selecting appropriate management practices.

Perhaps the most important management decision to make before implementing a reuse practice is crop selection. A list of other criteria that should be considered in the selection of crops for a reuse practice. In most cases, it is recommended that crops of high tolerance to salinity be selected when saline drainage water is to be used for irrigation. However, crops of intermediate tolerance (*e.g.* alfalfa, melons, tomatoes and wheat) may also be used in some cases, especially if the crop quality is sufficiently benefitted. For example, drainage water (EC 4-8 dS/m) significantly increased the protein content of wheat and alfalfa, soluble solids in melons and tomatoes, total digestible nutrients in alfalfa, and improved colour and netting of cantaloupe, and improved peelability in processing tomato. While improved plant quality should not be the major factor in adopting a reuse practice it may be an important factor in crop selection. Use of saline water to irrigate crops of intermediate tolerance to salinity is feasible, of course, only after seedlings have been established by good quality water.

Economics is also an important selection consideration, since it would be senseless to grow a high yielding crop without a marketable product and the potential for a positive cash flow. In the San Joaquin Valley in California, there is negative correlation between crop tolerance to salinity and economic value. It is unfortunate that there are not many crops that are both tolerant to salinity and have a high economic value. Asparagus is tolerant to salinity and has a high economic value, but harvesting is labour-intensive and costly.

The cyclic, "dual-rotation" reuse strategy, presupposes the availability of two water sources; the saline water to be utilized and the other a water of low salinity. Such reuse requires that the saline water be readily accessible for irrigation. Possible sources can be the drainage waters that are being discharged in pipes or canals from the irrigation project or that present in the underlying shallow groundwater system. Rainfall may also be the source of good-quality water, if it occurs at required times during the year to meet crop needs periodically and to leach excessive accumulations of soluble salts from the rootzone.

There are many different situations where the use of saline water for irrigation in the recommended strategy could be practical. One situation is where high quality water is available during the early growing season but is either too costly or too limited in supply to meet the entire seasons requirements. This situation is common in parts of India and Pakistan, for example. Where high-quality water costs are prohibitive, crops of moderate to high salt tolerance could be irrigated with saline drainage or groundwater, especially at later growth stages with economical advantage, even if this practice resulted in some reduction in yield relative to that obtainable with a full supply of fresh water. Use of saline water for irrigation reduces the amount of high-quality water required to grow crops and hence expands the water-resource base for crop production.

Another situation conducive for such reuse is one where drainage water disposal, or a means of lowering an excessively shallow water table, is impractical due to physical, environmental, social or political factors. Reuse of the drainage water for irrigation in this situation decreases the volume of drainage water requiring disposal or treatment, and the associated costs. Furthermore, a reduction in the drainage volume also reduces the salt loading of the receiving water. Many growers in the San Joaquin Valley of California are presently undertaking reuse of drainage water, at least as a temporary solution, in order to reduce drainage volume and to meet recently imposed discharge restrictions related to protection of the quality and ecology of receiving water systems.

A difficulty in adopting the cyclic, "dual-rotation" strategy may exist on small farms where the drainage water produced on-site is too little or does not coincide with peak crop-water demand. In the San Joaquin Valley in California, farms are often sufficiently large but peak drain water flow occurs from January to June when most crops would require high quality water. Sole use of drainage water later in the season may not be feasible if the flow rate needed for irrigation exceeds the flow rate from the drains. To avoid this limitation, surface storage reservoirs can be constructed to store the drainage water until its use is required. An option is to plug the subsurface drains and allow the soil to act as the reservoir. The latter option would not take land out of production for water storage purposes. However, regardless of where the drainage water is stored, a drainage water collection and irrigation system should be designed and operated with "reuse" in mind in order to implement this strategy most efficiently.

One method of collecting sufficient quantities of drainage water is to install a network of interceptor drains in areas with shallow water tables. A submersible pump could be placed in collector sumps as a means to access the drainage water. The size of the area that can be irrigated using such "drainage water" will vary, of course, depending on the capacity of the drainage system. To surface irrigate effectively, at least 10 litres/min/ha is required. Another way to collect drainage water is to install a network of shallow wells in strategic shallow water table areas. The wells can be connected to a common drainage manifold to facilitate collection and distribution. Consultation with irrigation and drainage engineers is advised before installing any drainage water collection system for irrigation use.

The long-term feasibility of using drainage water for irrigation in order to reduce drainage volume would likely be increased if implemented on a project or regional scale, rather than on a farm scale. Regional management permits reuse in dedicated areas so as to avoid the successive increase in concentration of the drainage water that would occur if the reuse process were to operate on the same water supply and same land area (*i.e.* in a closed loop). With regional management, certain areas in the region can be dedicated to reuse while other areas such as upslope areas, are irrigated solely with high quality water as usual. The second-generation drainage water from the primary reuse area is discharged to other dedicated reuse areas where even more salt-tolerant crops are grown, or to regional evaporation ponds or to treatment plants. Ideally, regional coordination and cost-sharing among growers should be undertaken in such a regional reuse system.

A novel means of "treating" saline waste waters before their ultimate disposal is to use them to irrigate specific crops that have the ability to accumulate large quantities of undesirable constituents (*e.g.* Se, Mo, NO_3, B, etc.) in the plants, in order to help reduce adverse ecological effects of disposal. The feasibility of biofiltration, the term used to describe this process, has been demonstrated by Cervinka *et al.* (1987), and Wu *et al.* (1987). They found that mustard, some grasses and certain native plant species found in California are effective in accumulating substantial amounts of Se in their shoots. This alternative "reuse" practice is most attractive where: (i) drainage disposal problems exist related to a potentially toxic trace constituent, (ii) a bioaccumulator with economic value exists, and (iii) other treatment processes are either unavailable or too expensive.

An alternative reuse strategy that is often advocated is to blend water supplies before or during irrigation. Blending may be appropriate provided the drainage or shallow groundwater is not too saline *per se* for the crop to be grown. However, in many cases this approach is inappropriate for the reasons. If the blending strategy is adopted, there must be a controlled means of mixing the water supplies. Shalhevet (1984) and Meiri *et al.* (1986) described two blending processes (i) network dilution, and (ii) soil dilution. With network dilution, water supplies are blended in the irrigation conveyance system. With soil dilution, the soil acts as the medium for mixing water of different qualities. A network blending system must be designed and installed if the blending strategy is to be adopted. The theory and design of dilution control systems and their use in irrigation networks has been developed by Sinai *et al.* (1985; 1989).

The cyclic strategy is preferred over the blending strategy in that (i) more salt sensitive crops can be included in the rotation, (ii) a blending facility is not required, and (iii) there is less danger in losing "usable water" for the crop. However, the cyclic strategy will require larger quantities of drainage water during the irrigations where it is used (since the water is not blended) and thus a storage system may be required in order to supply sufficient water for an effective irrigation. In summary, the "cyclic" strategy has more potential and flexibility than does the "blending" strategy, although the latter strategy is easier to implement in some cases.

Another concern besides excessive salinity build-up as regards the long-term feasibility of using saline water for irrigation is that of soil permeability and tilth. The likelihood of these problems increase as SAR increases and as electrical conductivity decreases. Therefore, adverse effects are most likely to occur during the periods of rainfall and irrigation using low-salinity water on soils previously irrigated with sodic, saline water. Such problems occurred at an experimental "reuse" site in California following pre-season rains and pre-irrigation with 0.3 dS/m canal water where sodic, saline water [9000 mg/l TDS; SAR = 30 $(mmol_c/l)^{1/2}$] had been used for irrigation for four consecutive years (Rolston *et al.* 1988). The consequence was impermeable, crusted soils and poor stand establishment. Whether such a problem will occur, or not, depends upon whether the EC of the high quality water is less than the threshold value, given the SAR of the saline water. Some combinations of the two waters are not permissible. This problem can often be controlled by the use of amendments and appropriate tillage.

Soil salinity under the cyclic strategy will fluctuate more, both spatially and temporally than in soils irrigated with conventional water supplies. Therefore, predicting plant response will be more difficult under these conditions. Hence, long-term effects on soil salination should be monitored using the techniques described earlier. Management must be adjusted to keep the average rootzone salinity levels within acceptable limits for the crop being grown, considering its stage of growth.

Many saline waters contain certain elements, such as boron and chloride, that can potentially accumulate in plants, especially woody, perennial ones, to levels that cause foliar injury and a subsequent reduction in their yield. In such cases, toxicity may produce more long-term detrimental effects than does salinity. Since boron is adsorbed by the soil it requires longer to build to toxic levels in the soil solution and it requires more leaching to remove its excessive accumulations than does salinity. Thus long-term accumulation in the soil of potential toxicants must be considered, since toxic effects may not become evident for years and may be more difficult to eliminate. Water containing excessive concentrations of B or Cl should not be used to irrigate perennial crops, since (i) these crops are generally more sensitive to specific-ion effects (ii) they represent a long-term investment, and (iii) they will have a long time opportunity to accumulate toxic levels. This same concern applies to growing perennial crops in the presence of a shallow groundwater that contains solutes potentially toxic to the plant.

Another consideration as regards use of saline water for irrigation is the potential of the plant to accumulate certain elements (such as Se, Mo, heavy metals) that are toxic to consumers of the crops (humans and animals). For example, in the San Joaquin Valley of California, drainage water in several locations contains unusually high levels of Se ([3] 50/m g Se/1). Although Se is essential to humans and animals in small amounts, excessive amounts can cause Se toxicosis.

In the San Joaquin Valley, melons and processing tomatoes irrigated with drainage water containing 250 to 350 m g Se/l accumulated elevated levels of Se in the fruit (250 to 750/m g/kg, dry wt.) that, while not an immediate health hazard, might become so to one whose diet was mostly restricted to such food. Many forages and native plant species have the potential to accumulate excessive amounts of Se. Since grazing animals consume larger quantities of plant mass than do humans, they have a greater potential for being "poisoned" in this manner.

Since plants vary in their ability to absorb and translocate toxic elements, crops that accumulate large quantities of toxic elements in the edible animal organs should also be avoided when using saline waters containing such elements for irrigation. Fleming (1962) found that Se concentrations were higher in *Cruciferae* (cabbage, cress, radish, rape and turnips), *Liliacea* (onion), and *Leguminosae* (clover and peas) than in *Compositae* (artichoke and lettuce), *Gramineae* (barley, oats, rye grass, and wheat), and *Umbelilferae* (parsnip and carrots) when grown on seleniferous soils in Ireland.

It is also important to understand how the toxic constituent is partitioned within the plant. In most annual fruit and vegetable crops selenium accumulates more in the leaves than in the fruit, but exceptions exist. If the saline water in question contains high levels of a potentially toxic element, the user should obtain expert advice.

FOOD PROCESSING INDUSTRIES

Food processing industries and practices include the following:

Canning

Canning is a method of preserving food in which the food is processed and sealed in an airtight container, providing a typical shelf life ranging from 1 year to 5 years and under specific circumstances a freeze dried canned product can last as long as 30 years and can still be safely consumed. The process was first developed as a French military discovery by Nicolas Appert in 1810. The packaging prevents microorganisms from entering and proliferating inside. To prevent the food from being spoiled before and during containment, quite a number of methods are used: pasteurisation, boiling (and other applications of high temperature over a period of time), refrigeration, freezing, drying, vacuum treatment, antimicrobial agents that are natural to the recipe of the foods being preserved, a sufficient dose of ionizing radiation, submersion in a strong saline solution, acid, base, osmotically extreme (for example very sugary) or other microbe-challenging environments.

Other than sterilization, no method is perfectly dependable as a preservative. For example, the microorganism *Clostridium botulinum* (which causes botulism), can only be eliminated at temperatures above the boiling point. From a public safety point of view, foods with low acidity (a pH more than 4.6) need sterilization under high temperature (116-130 °C). To achieve temperatures above the boiling point requires the use of a pressure canner. Foods that must be pressure canned include most vegetables, meats, seafood, poultry, and dairy products. The only foods that may be safely canned in an ordinary boiling water bath are highly acidic ones with a pH below 4.6, such as fruits, pickled vegetables, or other foods to which acidic additives have been added.

History and Development of Canning

During the first years of the Napoleonic Wars, the French government offered a hefty cash award of 12,000 francs to any inventor who could devise a cheap and effective method of preserving large amounts of food. The larger armies of the period required increased, regular supplies of quality food. Limited food availability was among the factors limiting military campaigns to the summer and autumn months.

In 1809, a French confectioner and brewer, Nicolas Appert, observed that food cooked inside a jar did not spoil unless the seals leaked, and developed a method of sealing food in glass jars. The reason for lack of spoilage was unknown at the time, since it would be another 50 years before Louis Pasteur demonstrated the role of microbes in food spoilage. However, glass containers presented challenges for transportation. Glass jars were largely replaced in commercial canneries with cylindrical tin or wrought-iron canisters (later shortened to "cans") following the work of Peter Durand (1810). Cans are cheaper and quicker to make, and much less fragile than glass jars. Glass jars have remained popular for some high-value products and in home canning. Can openers were not invented for another thirty years — at first, soldiers had to cut the cans open with bayonets or smash them open with rocks. The French Army began experimenting with issuing canned foods to its soldiers, but the slow process of canning foods and the even slower development and transport stages prevented the army from shipping large amounts across the French Empire, and the war ended before the process was perfected. Unfortunately for Appert, the factory which he had built with his prize money was razed in 1814 by Allied soldiers invading France.

Following the end of the Napoleonic Wars, the canning process was gradually employed in other European countries and in the US. Based on Appert's methods of food preservation, Peter Durand patented a process in the United Kingdom in 1810.

He did not develop the process, selling his patent in 1811 to Bryan Donkin and John Hall, who were in business as Donkin Hall and Gamble, of Bermondsey. Bryan Donkin developed the process of packaging food in sealed airtight cans, made of tinned wrought iron. Initially, the canning process was slow and labour-intensive, as each large can had to be hand-made, and took up to six hours to cook, making canned food too expensive for ordinary people. The main market for the food at this stage was the British Army and Royal Navy. By 1817 Donkin recorded that he had sold £3000 worth of canned meat in six months. In 1824 Sir William Edward Parry took canned beef and pea soup with him on his voyage to the Arctic in HMS Fury, during his search for a northwestern passage to India. In 1829, Admiral Sir James Ross also took canned food to the Arctic, as did Sir John Franklin in 1845. Some of his stores were found by the search expedition led by Captain (later Admiral Sir) Leopold McLintock in 1857. One of these cans was opened in 1939, and was edible and nutritious, though it was not analysed for contamination by the lead solder used in its manufacture.

Throughout the mid-19th century, canned food became a status symbol amongst middle-class households in Europe, becoming something of a frivolous novelty. Early methods of manufacture employed poisonous lead solder for sealing the cans, which may have worsened the disastrous outcome of the 1845 Franklin expedition to chart and navigate the Northwest Passage.

Increasing mechanisation of the canning process, coupled with a huge increase in urban populations across Europe, resulted in a rising demand for canned food. A number of inventions and improvements followed, and by the 1860s smaller machine-made steel cans were possible, and the time to cook food in sealed cans had been reduced from around six hours to thirty minutes. Canned food also began to spread beyond Europe — Robert Ayars established the first American canning factory in New York City in 1812, using improved tin-plated wrought-iron cans for preserving oysters, meats, fruits and vegetables. Demand for canned food greatly increased during wars. Large-scale wars in the nineteenth century, such as the Crimean War, American Civil War, and Franco-Prussian War introduced increasing numbers of working-class men to canned food, and allowed canning companies to expand their businesses to meet military demands for non-perishable food, allowing companies to manufacture in bulk and sell to wider civilian markets after wars ended. Urban populations in Victorian era Britain demanded ever-increasing quantities of cheap, varied, quality food that they could keep at home without having to go shopping daily.

In response, companies such as Nestlé, Heinz, and others emerged to provide quality canned food for sale to working class city-dwellers. In particular, Crosse and Blackwell took over the concern of Donkin Hall and Gamble. The late 19th century saw the range of canned food available to urban populations greatly increase, as canners competed with each other using novel foodstuffs, highly decorated printed labels, and lower prices.

Demand for canned food skyrocketed during World War I, as military commanders sought vast quantities of cheap, high-calorie food to feed their millions of soldiers, which could be transported safely, survive trench conditions, and not spoil in transport. Throughout the war, soldiers generally subsisted on low-quality canned foodstuffs, such as the British "Bully Beef" (cheap corned beef), pork and beans and Maconochies Irish Stew, but by 1916 widespread boredom with cheap canned food amongst soldiers resulted in militaries purchasing better-quality food to improve morale, and the complete meals in a can began to appear. In 1917 the French Army began issuing canned French cuisine, such as coq au vin, while the Italian Army experimented with canned ravioli and spaghetti bolognese.

Shortages of canned food in the British Army in 1917 led to the government issuing cigarettes and amphetamines to soldiers to suppress their appetites. After the war, companies that had supplied military canned food improved the quality of their goods for civilian sale. Today, tin-coated steel is the material most commonly used. Laminate vacuum pouches are also used for canning, such as used in MREs and Capri Sun drinks.

Double Seams

Invented in 1888 by Max Ams, modern double seams provide an airtight seal to the tin can. This airtight nature is crucial to keeping bacteria out of the can and keeping its contents sealed inside. Thus, double seamed cans are also known as Sanitary Cans. Developed in 1900 in Europe, this sort of can was made of the traditional cylindrical body made with tin plate. The two ends (lids) were attached using what is now called a double seam. A can thus sealed is impervious to the contamination by creating two tight continuous folds between the can's cylindrical body and the lids. This eliminated the need for solder and allowed improvements in manufacturing speed, reducing cost. Double seaming uses rollers to shape the can, lid and the final double seam.

To make a sanitary can and lid suitable for double seaming, manufacture begins with a sheet of coated tin plate. To create the can body, rectangles are cut and curled around a die, and welded together creating a cylinder with a side seam. Rollers are then used to flare out one or both ends of the cylinder to create a quarter circle flange around the circumference. Precision is required to ensure that the welded sides are perfectly aligned, as any misalignment will cause inconsistent flange shape, compromising its integrity. A circle is then cut from the sheet using a die cutter. The circle is shaped in a stamping press to create a downward countersink to fit snugly in to the can body. The result can be compared to an upside down and very flat top hat. The outer edge is then curled down and around about 140 degrees using rollers to create the end curl. The result is a steel tube with a flanged edge, and a countersunk steel disc with a curled edge. A rubber compound is put inside the curl.

Seaming

The body and end are brought together in a seamer and held in place by the base plate and chuck, respectively. The base plate provides a sure footing for the can body during the seaming operation and the chuck fits snugly in to the end (lid). The result is the countersink of the end sits inside the top of the can body just below the flange. The end curl protrudes slightly beyond the flange.

First Operation

Once brought together in the seamer, the seaming head presses a first operation roller against the end curl. The end curl is pressed against the flange curling it in toward the body and under the flange. The flange is also bent downward, and the end and body are now loosely joined together. The first operation roller is then retracted. At this point five thicknesses of steel exist in the seam. From the outside in they are:

- End
- Flange
- End Curl
- Body
- Counters.

Second Operation

The seaming head then engages the second operation roller against the partly formed seam. The second operation presses all five steel components together tightly to form the final seal. The five layers in the final seam are then called; a) End, b) Body Hook, c) Cover Hook, d) Body, e) Countersink. All sanitary cans require a filling medium within the seam because otherwise the metal-to-metal contact will not maintain a hermetic seal. In most cases, a rubberized compound is placed inside the end curl radius, forming the critical seal between the end and the body. Probably the most important innovation since the introduction of double seams is the welded side seam. Prior to the welded side seam, the can body was folded and/or soldered together, leaving a relatively thick side seam. The thick side seam required that the side seam end juncture at the end curl to have more metal to curl around before closing in behind the Body Hook or flange, with a greater opportunity for error.

Seamer Setup and Quality Assurance

Many different parts during the seaming process are critical in ensuring that a can is airtight and vacuum sealed. The dangers of a can that is not hermetically sealed are contamination by foreign objects (bacteria or fungicide sprays), or that the can could leak or spoil. One important part is the seamer setup. This process is usually performed by an experienced technician. Amongst the parts that need setup are seamer rolls and chucks which have to be set in their exact position (using a feeler gauge or a clearance gauge). The lifter pressure and position, roll and chuck designs, tooling wear, and bearing wear all contribute to a good double seam. Incorrect setups can be non-intuitive. For example, due to the springback effect, a seam can appear loose, when in reality it was closed too tight and has opened up like a spring.

For this reason, experienced operators and good seamer setup are critical to ensure that double seams are properly closed. Quality control usually involves taking full cans from the line - one per seamer head, at least once or twice per shift, and performing a teardown operation (wrinkle/tightness), mechanical tests (external thickness, seamer length/height and countersink) as well as cutting the seam open with a twin blade saw and measuring with a double seam inspection system. The combination of these measurements will determine the seam's quality. Use of a [Statistical Process Control] or [SPC] software in conjunction with a manual double seam monitor, computerized double seam scanner, or even a fully-automatic double seam inspection system makes the laborious process of double seam inspection faster and much more accurate. Statistically tracking the performance of each head or seaming station of the [can seamer] allows for better prediction of can seamer issues, and may be used to plan maintenance when convenient: rather that to simply react after bad or unsafe cans have been produced.

Nutrition Value

Canning is a way of processing food to extend its shelf life. The idea is to make food available and edible long after the processing time. A 1997 study found that canned fruits and vegetables provide as much dietary fiber and vitamins as the same corresponding fresh or frozen foods, and in some cases, even more. The heating process during canning appears to make dietary fiber more soluble, and therefore more readily fermented in the colon into gases and physiologically active byproducts. Canned tomatoes have a higher available lycopene content.

POTENTIAL HAZARDS

To improve food safety for those who eat canned food, governments have enacted laws requiring alphanumeric codes being put on food cans during manufacture indicating information relevant to health, such as the date of canning, etc.

Migration of Can Components

In canning toxicology, *migration* is the movement of substances from the can itself into the contents. Potential toxic substances that can migrate are lead, causing lead poisoning, or bisphenol A, a potential endocrine disruptor that is an ingredient in the epoxy commonly used to coat the inner surface of cans.

Salt Content

Canned food can be a major source of dietary salt (sodium chloride). Too much salt increases the risk of health problems, including high blood pressure. Therefore, health authorities have recommended limitations of dietary sodium. Many canned products are available in low-salt and no-salt alternatives.

Botulism

Foodborne botulism results from contaminated foodstuffs in which *C. botulinum* spores have been allowed to germinate and produce botulism toxin, and this typically occurs in canned non-acidic food substances. *C. botulinum* prefers low oxygen environments, and can therefore grow in canned foods. Botulism is a rare but serious paralytic illness, leading to paralysis that typically starts with the muscles of the face and then spreads towards the limbs.

In severe forms, it leads to paralysis of the breathing muscles and causes respiratory failure. In view of this life-threatening complication, all suspected cases of botulism are treated as medical emergencies, and public health officials are usually involved to prevent further cases from the same source.

Canning and The Recession

Canned Goods (and canning supplies) sell particularly well in times of recession due to the tendency of financially-stressed individuals to engage in cocooning, a term used by retail analysts to describe the phenomenon in which people actively avoid straying from their houses. In February 2009, the recession-laden United States saw an 11.5% rise in sales of canning-related items.

Food Marketing

Food marketing brings together the producer and the consumer. It is the chain of activities that brings food from "farm gate to plate." The marketing of even a single food product can be a complicated process involving many producers and companies. For example, fifty-six companies are involved in making one can of chicken noodle soup. These businesses include not only chicken and vegetable processors but also the companies that transport the ingredients and those who print labels and manufacture cans. The food marketing system is the largest direct and indirect nongovernment employer in the United States.

THE THREE HISTORICAL PHASES OF FOOD MARKETING

There are three historical phases of food marketing: the fragmentation phase (before 1870–1880), the unification phase (1880–1950), and the segmentation phase (1950 and later). In the fragmentation phase, the United States was divided into numerous geographic fragments because transporting food was expensive, leaving most production, distribution, and selling locally based.

In the unification phase, distribution was made possible by railroads, coordination of sales forces was made possible by the telegraph and telephone, and product consistency was made possible by advances in manufacturing. This new distribution system was led by meat processors such as Armour and Swift in midwestern cities and by companies such as Heinz, Quaker Oats, Campbell Soup, and Coca-Cola, which sold their brands nationally. Advertising in print media and direct marketing through demonstrations at stores and public venues were among the prime marketing tools. The initial Crisco campaign, in 1911, was an example. In the segmentation phase (1950 and later) radio, television and internet advertising made it possible for a wider range of competing products to focus on different benefits and images and thus appeal to different demographic and psychographic markets. Distribution via the new national road system strengthened national brands.

The Food Marketing Mix and the Four Ps of Marketing

The four components of food marketing are often called the "four Ps" of the marketing mix because they relate to product, price, promotion, and place. One reason food manufacturers receive the largest percentage of the retail food dollar is that they provide the most differentiating, value-added service. The money that manufacturers invest in developing, pricing, promotion, and placing their products helps differentiate a food product on the basis of both quality and brand-name recognition.

Product

In deciding what type of new food products a consumer would most prefer, a manufacturer can either try to develop a new food product or try to modify or extend an existing food. For example, a sweet, flavored yogurt drink would be a new product, but milk in a new flavor (such as chocolate strawberry) would be an extension of an existing product. There are three steps to both developing and extending: generate ideas, screen ideas for feasibility, and test ideas for appeal. Only after these steps will a food product make it to national market. Of one hundred new food product ideas that are considered, only six make it to a supermarket shelf.

The food industry faces numerous marketing decisions. Money can be invested in brand building (through advertising and other forms of promotion) to increase either quantity demanded or the price consumers are willing to pay for a product. Coca Cola, for example, spends a great deal of money both on perfecting its formula and on promoting the brand. This allows Coke to charge more for its product than can makers of regional and smaller brands. Manufacturers may be able to leverage their existing brand names by developing new product lines. For example, Heinz started out as a brand for pickles but branched out into ketchup. Some brand extensions may involve a risk of damage to the original brand if the quality is not good enough. Coca Cola, for example, refused to apply the Coke name to a diet drink back when artificial sweeteners had a significantly less attractive taste. Coke created Tab Cola, but only when aspartame (NutraSweet) was approved for use in soft drinks did Coca Cola come out with a Diet Coke. Manufacturers that have invested a great deal of money in brands may have developed a certain level of consumer brand loyalty—that is, a tendency for consumers to continue to buy a preferred brand even when an attractive offer is made by competitors. For loyalty to be present, it is not enough to merely observe that the consumer buys the same brand consistently. The consumer, to be brand loyal, must be able to actively resist promotional efforts by competitors. A brand loyal consumer will continue to buy the preferred brand even if a competing product is improved, offers a price promotion or premium, or receives preferred display space. Some consumers have multi-brand loyalty. Here, a consumer switches between a few preferred brands. The consumer may either alternate for variety or may, as a rule of thumb, buy whichever one of the preferred brands is on sale. This consumer, however, would not switch to other brands on sale. Brand loyalty is, of course, a matter of degree. Some consumers will not switch for a moderate discount, but would switch for a large one or will occasionally buy another brand for convenience or variety.

Price

In profitably pricing the food, the manufacturer must keep in mind that the retailer adds approximately 50 percent to the price of a wholesale product. For example, a frozen food sold in a retail store for $4.50 generates an income of $3.00 for the manufacturer. This money has to pay for the cost of producing, packaging, shipping, storing, and selling the product.

Promotion

Promoting a food to consumers is done out of store, in store, and on package. Advertisements on television and in magazines are attempts to persuade consumers to think favourably about a product, so that they go to the store to purchase the product. In addition to advertising, promotions can also include Sunday newspaper ads that offer coupons such as cents-off and buy-one-get-one-free offers.

Place

Place refers to the distribution and warehousing efforts necessary to move a food from the manufacturer to a location where a consumer can buy it. It can also relate to the place within a store that it is located in a place like Sobey's, IGA, Safeway, Walmart and many other gargantuious corporations that have yogurt in stock. The food marketing system in the United States is an amazingly flexible one. Consumer focus helps marketers anticipate the demands of consumers, and production focus helps them respond to changes in the market. The result is a system that meets the ever-changing demands of consumers.

PACKAGING AND LABELLING

Packaging is the science, art and technology of enclosing or protecting products for distribution, storage, sale, and use. Packaging also refers to the *process* of design, evaluation, and production of packages. Packaging can be described as a *coordinated system* of preparing goods for transport, warehousing, logistics, sale, and end use. Packaging contains, protects, preserves, transports, informs, and sells. In many countries it is fully integrated into government, business, institutional, industrial, and personal use. Package labelling (en-GB) or labelling (en-US) is any written, electronic, or graphic communications on the packaging or on a separate but associated label.

The first packages used the natural materials available at the time: Baskets of reeds, wineskins (Bota bags), wooden boxes, pottery vases, ceramic amphorae, wooden barrels, woven bags, etc. Processed materials were used to form packages as they were developed: for example, early glass and bronze vessels. The study of old packages is an important aspect of archaeology.

The earliest recorded use of paper for packaging dates back to 1035, when a Persian traveler visiting markets in Cairo noted that vegetables, spices and hardware were wrapped in paper for the customers after they were sold. Iron and tin plated steel were used to make cans in the early 19th century. Paperboard cartons and corrugated fiberboard boxes were first introduced in the late 19th century.

Packaging advancements in the early 20th century included Bakelite closures on bottles, transparent cellophane overwraps and panels on cartons, increased processing efficiency and improved food safety. As additional materials such as aluminum and several types of plastic were developed, they were incorporated into packages to improve performance and functionality. In-plant recycling has long been common for production of packaging materials. Post-consumer recycling of aluminum and paper based products has been economical for many years: since the 1980s, post-consumer recycling has increased due to curbside recycling, consumer awareness, and regulatory pressure. As of 2003, the packaging sector accounted for about two percent of the gross national product in developed countries. About half of this market was related to food packaging.

THE PURPOSES OF PACKAGING AND PACKAGE LABELS

Packaging and package labelling have several objectives

- Physical protection - The objects enclosed in the package may require protection from, among other things, mechanical shock, vibration, electrostatic discharge, compression, temperature, etc.
- Barrier protection - A barrier from oxygen, water vapor, dust, etc., is often required. Permeation is a critical factor in design. Some packages contain desiccants or Oxygen absorbers to help extend shelf life. Modified atmospheres or controlled atmospheres are also maintained in some food packages. Keeping the contents clean, fresh, sterile and safe for the intended shelf life is a primary function.
- Containment or agglomeration - Small objects are typically grouped together in one package for reasons of efficiency. For example, a single box of 1000 pencils requires less physical handling than 1000 single pencils. Liquids, powders, and granular materials need containment.
- Information transmission - Packages and labels communicate how to use, transport, recycle, or dispose of the package or product. With pharmaceuticals, food, medical, and chemical products, some types of information are required by governments. Some packages and labels also are used for track and trace purposes.
- Marketing - The packaging and labels can be used by marketers to

encourage potential buyers to purchase the product. Package graphic design and physical design have been important and constantly evolving phenomenon for several decades. Marketing communications and graphic design are applied to the surface of the package and (in many cases) the point of sale display.

- Security - Packaging can play an important role in reducing the security risks of shipment. Packages can be made with improved tamper resistance to deter tampering and also can have tamper-evident features to help indicate tampering. Packages can be engineered to help reduce the risks of package pilferage: Some package constructions are more resistant to pilferage and some have pilfer indicating seals. Packages may include authentication seals and use security printing to help indicate that the package and contents are not counterfeit. Packages also can include anti-theft devices, such as dye-packs, RFID tags, or electronic article surveillance tags that can be activated or detected by devices at exit points and require specialized tools to deactivate. Using packaging in this way is a means of loss prevention.
- Convenience - Packages can have features that add convenience in distribution, handling, stacking, display, sale, opening, reclosing, use, dispensing, and reuse.
- Portion control - Single serving or single dosage packaging has a precise amount of contents to control usage. Bulk commodities (such as salt) can be divided into packages that are a more suitable size for individual households. It is also aids the control of inventory: selling sealed one-liter-bottles of milk, rather than having people bring their own bottles to fill themselves.

PACKAGING TYPES

Packaging may be looked at as being of several different types. For example a transport package or distribution package can be the shipping container used to ship, store, and handle the product or inner packages. Some identify a consumer package as one which is directed toward a consumer or household.

Packaging may be described in relation to the type of product being packaged: medical device packaging, bulk chemical packaging, over-the-counter drug packaging, retail food packaging, military materiel packaging, pharmaceutical packaging, etc. It is sometimes convenient to categorize packages by layer or function: "primary", "secondary", etc.

- Primary packaging is the material that first envelops the product and holds it. This usually is the smallest unit of distribution or use and is the package which is in direct contact with the contents.

- Secondary packaging is outside the primary packaging, perhaps used to group primary packages together.
- Tertiary packaging is used for bulk handling, warehouse storage and transport shipping. The most common form is a palletized unit load that packs tightly into containers.

These broad categories can be somewhat arbitrary. For example, depending on the use, a shrink wrap can be primary packaging when applied directly to the product, secondary packaging when combining smaller packages, and tertiary packaging on some distribution packs.

SYMBOLS USED ON PACKAGES AND LABELS

Many types of symbols for package labelling are nationally and internationally standardized. For consumer packaging, symbols exist for product certifications, trademarks, proof of purchase, etc. Some requirements and symbols exist to communicate aspects of consumer use and safety, for example the estimated sign that notes conformance to EU weights and measures accuracy regulations.

Examples of environmental and recycling symbols include the recycling symbol, the resin identification code and the "Green Dot". Bar codes, Universal Product Codes, and RFID labels are common to allow automated information management in logistics and retailing. Country of Origin Labelling is often used.

Shipping Container Labelling

Technologies related to shipping containers are identification codes, bar codes, and electronic data interchange (EDI). These three core technologies serve to enable the business functions in the process of shipping containers throughout the distribution channel.

Each has an essential function: identification codes either relate product information or serve as keys to other data, bar codes allow for the automated input of identification codes and other data, and EDI moves data between trading partners within the distribution channel. Elements of these core technologies include UPC and EAN item identification codes, the SCC-14 (UPC shipping container code), the SSCC-18 (Serial Shipping Container Codes), Interleaved 2-of-5 and UCC/EAN-128 (newly designated GS1-128) bar code symbologies, and ANSI ASC X12 and UN/EDIFACT EDI standards.

Small parcel carriers often have their own formats. For example, United Parcel Service has a MaxiCode 2-D code for parcel tracking. RFID labels for shipping containers are also increasing in usage. A Wal-Mart division, Sam's Club, has also moved in this direction and is putting pressure on its suppliers for compliance. Shipments of hazardous materials or dangerous goods have special information and symbols (labels, plackards, etc.) as required by UN, country, and specific carrier requirements.

Package Development Considerations

Package design and development are often thought of as an integral part of the new product development process. Alternatively, development of a package (or component) can be a separate process, but must be linked closely with the product to be packaged. Package design starts with the identification of all the requirements: structural design, marketing, shelf life, quality assurance, logistics, legal, regulatory, graphic design, end-use, environmental, etc.

The design criteria, performance (specified by package testing), completion time targets, resources, and cost constraints need to be established and agreed upon. An example of how package design is affected by other factors is the relationship to logistics. When the distribution system includes individual shipments by a small parcel carrier, the sortation, handling, and mixed stacking make severe demands on the strength and protective ability of the transport package.

If the logistics system consists of uniform palletized unit loads, the structural design of the package can be designed to those specific needs: vertical stacking, perhaps for a longer time frame. A package designed for one mode of shipment may not be suited for another.

With some types of products, the design process involves detailed regulatory requirements for the package. For example with packaging foods, any package components that may contact the food are food contact materials. Toxicologists and food scientists need to verify that the packaging materials are allowed by applicable regulations. Packaging engineers need to verify that the completed package will keep the product safe for its intended shelf life with normal usage. Packaging processes, labelling, distribution, and sale need to be validated to comply with regulations and have the well being of the consumer in mind.

Sometimes the objectives of package development seem contradictory. For example, regulations for an over-the-counter drug might require the package to be tamper-evident and child resistant: These intentionally make the package difficult to open. The intended consumer, however, might be handicapped or elderly and be unable to readily open the package. Meeting all goals is a challenge. Package design may take place within a company or with various degrees of external packaging engineering: independent contractors, consultants, vendor evaluations, independent laboratories, contract packagers, total outsourcing, etc. Some sort of formal Project planning and Project management methodology is required for all but the simplest package design and development programmes. An effective quality management system and Verification and Validation protocols are mandatory for some types of packaging and recommended for all.

Environmental Considerations

Package development involves considerations for sustainability, environmental responsibility, and applicable environmental and recycling regulations.

It may involve a life cycle assessment which considers the material and energy inputs and outputs to the package, the packaged product (contents), the packaging process, the logistics system, waste management, etc. It is necessary to know the relevant regulatory requirements for point of manufacture, sale, and use.

The traditional "three R's" of reduce, reuse, and recycle are part of a waste hierarchy which may be considered in product and package development.

- Minimization – (also "source reduction") The mass and volume of packaging (per unit of contents) can be measured and used as one of the criteria to minimize during the package design process. Usually "reduced" packaging also helps minimize costs. Packaging engineers continue to work toward reduced packaging.
- Prevention – Waste prevention is a primary goal. Packaging should be used only where needed. Proper packaging can also help prevent waste. Packaging plays an important part in preventing loss or damage to the packaged-product (contents). Usually, the energy content and material usage of the product being packaged are much greater than that of the package. A vital function of the package is to protect the product for its intended use: if the product is damaged or degraded, its entire energy and material content may be lost.
- Reuse – The reuse of a package or component for other purposes is encouraged. Returnable packaging has long been useful (and economically viable) for closed loop logistics systems. Inspection, cleaning, repair and recouperage are often needed. Some manufacturers re-use the packaging of the incoming parts for a product, either as packaging for the outgoing product or as part of the product itself.
- Recycling – Recycling is the reprocessing of materials (pre- and post-consumer) into new products. Emphasis is focused on recycling the largest primary components of a package: steel, aluminum, papers, plastics, etc. Small components can be chosen which are not difficult to separate and do not contaminate recycling operations.
- Energy recovery – Waste-to-energy and Refuse-derived fuel in

approved facilities are able to make use of the heat available from the packaging components.

- Disposal – Incineration, and placement in a sanitary landfill are needed for some materials. Certain states within the US regulate packages for toxic contents, which have the potential to contaminate emissions and ash from incineration and leachate from landfill. Packages should not be littered.

Development of sustainable packaging is an area of considerable interest by standards organizations, government, consumers, packagers, and retailers.

4

Promoting Food Security: The Indian Challenge

SUSTAINABLE FOOD SECURITY: FOOD SECURITY, INCOME AND POVERTY

As with 'sustainable development' there is no precise, universally accepted definition of 'food security'. In broad principle a state of food security exists where all individuals, families and communities enjoy consistent access to foods that in quantity, quality and biochemical composition provide hygienic, nutritional adequacy. The Brundtland Panel states: "Food security requires secure ownership of, or access to, food resources and income earning activities, including reserves and assets to offset risks, to ease shocks and meet all contingencies". This may be interpreted as all people having access to stocks and flows of food and cash sufficient to satisfy their basic nutritional needs. Everyone in the world need not be a subsistence farmer, but those who do not produce foods in quantities sufficient to satisfy their families' needs, must possess the means to acquire foods nutritionally adequate, readily accessible and affordable.

The report of the 1992 FAO/WHO international conference on nutrition states: "Food security is defined in its most basic form as access by all people at all times to the food needed for a healthy life. Food security has three dimensions: assurance of a safe, nutritionally adequate food supply at national and household levels; a reasonable degree of stability in the supply of food throughout each year and from one year to the next; every household having physical, social and economic access to food sufficient to meet its needs."

While there is agreement in broad principle on requirements for food security, concepts diverge significantly in semasiological and biological specifics, they change over time, and differ among specialists and agencies as to what constitutes a nutritionally adequate diet. Non-scientific and quasi-scientific publications frequently fail to make clear distinction among 'hunger', 'malnutrition', 'famine' and 'starvation'; between 'chronic' and 'transitory hunger'. The noun 'security' derives from the Latin '*securus*' which means 'safe', 'free from worry and care'. People blessed with food security are safe from the risk of famine, starvation, chronic hunger and malnutrition. 'Famine' describes extreme food scarcity among large populations.

'Starvation' is derived from an Old English word meaning to die a slow lingering death from cold, disease, insufficient food and/or water. 'Malnutrition' defines a significant insufficiency or imbalance in essential nutrient intake, 'sub-nutrition' sometimes describes people who have too little food. 'Hunger' describes distressing discomfort caused by want of food. 'Transitory hunger' is experienced from time to time by active persons who enjoy healthy appetites. 'Chronic hunger' describes a painful and debilitating condition caused by long deprivation of food sufficient to sustain healthy growth and activity.

The 1960s Freedom From Hunger Campaign, the 1974 World Congress' hope of eradicating hunger by 1984, and the FAO Millennium objective of reducing incidence of hunger by 50 per cent between 1990 and 2015, all relate to alleviation of chronic hunger and malnutrition. Roughly 18 per cent of the world's population (800 million, including 200m children), 25 per cent of South Asians (300 million people) and 35 per cent of sub-Saharan Africans (190 million) are classified by UN agencies as undernourished: their caloric intake is lower than is required for healthy activity. In 1980 30 per cent of all people in LICs were considered undernourished. By the year 2000 this proportion had fallen to 18 per cent.

In simplest terms, food security depends upon a dynamic balance among disposable income, demand, supply and distribution. Production of food crops, livestock and fisheries must progressively increase to satisfy expanding and diversifying needs and demands of growing populations. Efficient post-production systems, to ensure safe and effective protection, preservation, transformation and distribution must be economically integrated with production systems. During the early 1960s, affluent OECD nations were home to 34 per cent of world population and controlled close to 90 per cent of global GNP. By 1990, the affluent minority had fallen to 16 per cent of the total but controlled more than 82 per cent of global GNP. Between 1960 and 1990 a new group of middle-income newly industrialised countries emerged, most in East Asia and Latin America, a change with considerable implications for the Planet's economy and future patterns of food demand.

In 1990 Asians represented 52 per cent, in 2002 some 59 per cent of world population. Among Asian nations, middle-income communities are growing most rapidly, their diets diversifying with dilating demand for livestock products. Food insecurity, chronic hunger and under-nutrition are both dismal consequences and clear indicators of extreme poverty. People with money, power and control over negotiable assets rarely suffer from chronic hunger. Government politicians and military generals who suffer from chronic hunger are not a common spectacle around the world. Alleviation of food insecurity requires greater opportunities for paid employment and more equitable access to critical resources and assets among and within nations. A respected observer writes: "In the 1950s... it was assumed that adequate agricultural production would assure access to adequate food in the market and in the household. In the 1970s it became clear that production and availability alone do not lead to food security, since those who lack purchasing power (from permanent employment income) lack access to a balanced diet. Now it is evident that where availability and access are satisfactory, necessary biological absorption of food requires clean drinking water, environmental hygiene, primary health care and education".

Based on these considerations, the M. S. Swaminathan Research Foundation and the UN World Food Programme have published a *Food Insecurity Atlas for Rural India*. The atlas illustrates marked variability among Indian States in food availability, absorption of nutrients, vulnerability to chronic hunger and in sustainable agricultural production. The atlas describes such non-food factors as income earning opportunities, health care facilities, education, sanitation and environmental hygiene.

Due to climatic variability, years with too little or too much rain, damage and destruction by pests, parasites and pathogens, crop and livestock production inevitably change from year to year. To compensate for years of poor harvests governments and/or regional organisations should maintain reserve stocks to carry over surpluses from regions and seasons of abundance to those of scarcity. The cost of holding reserve stocks should be calculated as an insurance investment not an unwarranted expense.

Food security need not be synonymous with food self-sufficiency, a nation's ability to produce all food required by its population. For diverse reasons: climate, topography, soil and climatic conditions, many nations cannot be self-sufficient and must depend on food aid or imports.

Substantial reductions in food spoilage after harvest could be realised by establishing primary food preservation and protection facilities in rural areas where crops are cultivated and animals husbanded. Food saved from spoilage increases total food available and enhances food availability and security.

PUBLIC POLICY AND FOOD SECURITY

Estimates of 'Global food security', based on the quotient of total possible food production divided by world population, may be a diverting statistical exercise, but is of little practical value. Equitable access to the planet's resources, though highly desirable, never has been nor is ever likely to be a realisable objective. To state that over the next half-century world food production must exceed all that was produced since settled farming began 10,000 years ago is interesting but offers little practical guidance for long-term planning by national governments. Food security can only be assessed and assured for discrete nations, communities and families within communities, a process that is the responsibility of national and local governments. International agencies can help weak and inexperienced governments by defining criteria to be considered and by providing long-term development assistance.

During the 1930s, the League of Nations Committee on Nutrition reported: "In the past, movement towards better nutrition has been largely the result of unconscious, instinctive groping for a better, more abundant life. Now needed is planning for better nutrition policies based on (a) consumption: bringing essential food within reach of all people; (b) adequate supply: production and equitable distribution of all essential food sources". The report urges (a) recognition of nutrition policy as of primary national importance; (b) better, widespread education on human nutrition; (c) more equitable distribution of income: "it is the poorest people who are most nutritionally deprived".

A committee of the US National Research Council appears to agree: "An important cause of malnutrition is absence of policies and programmes that foster the best use of available food supplies... Governments continually make decisions that affect nutritional status with little knowledge of nutritional consequences. Alleviation of hunger and malnutrition depend upon increasing the right kinds of food, reducing poverty and improving the stability of food supplies". Publications that discuss food security are many, vast and voluminous. They suggest primary causes, and propose various courses of remedial action. The League of Nations reports are among many that insist that food security be of highest priority in all governmental policies. Reutlinger [1987] states that food insecurity results from inappropriate macro-economic policies, from local economic and political structures that restrict the ability of households and individuals to have access to foods sufficient to satisfy their needs. Food insecurity may be transitory, lasting for relatively short periods, or chronic and of long duration, a consequence of poor planning to ensure resources adequate to produce or purchase sufficient food. Transitory hunger frequently results from unstable food production systems, sharp rises in food prices, or depression of disposable family incomes. If extreme and extended, transitory hunger can exacerbate into chronic hunger and eventually into famine. Famines have ravaged even when food stocks were plentiful: the 1943 famine in Bengal was precipitated by the price of a limited food supply rising to levels unaffordable by poor people.

Sen [1987] relates food security to entitlements and ownerships; families and individuals must have adequate access and purchasing power. Food security depends not on estimates of national aggregate demand but on fair exchange in international trade of goods and services. The poor must inevitably rely on food aid provided by social programmes. Chronic hunger and malnutrition are less influenced by a nation's total food supply than the degree of equity with which available resources are distributed and made accessible. Food and agricultural policies to ensure food security for all citizens should be a high priority for every government. Malnourished people, too weak to work effectively, eventually impose a financial burden on the nation's economy. Though some observers argue that food stocks of grain are of value only to alleviate temporary food insecurity, the prime purpose of strategic food stocks is to provide emergency reserves against regional, national, local and seasonal shortages, to stabilise food prices and counteract fluctuations in supply and availability. Reserve food stocks are customarily controlled by government agencies, authorised to maintain quantities sufficient for defined time periods. Reserve stocks must be replenished when they fall below a prescribed safe minimum. Reserve stocks, essential in regions susceptible to frequent drought, provide short-term relief but cannot be considered a panacea for chronic food insecurity.

Santorum and Gray [1993] list factors to be weighed in determining minimum reserve stocks. Stocks stored on farms are generally sufficient only to satisfy farmers' needs from one harvest to the next. Countries within a region that cannot individually afford to maintain national reserves can collectively contribute to and draw upon strategically dispersed food stocks. During the 1980s the Southern African Development Community (SADC) created and maintained a regional food security programme with grain stocks upon which member nations could draw in time of need. Where many are too poor to buy adequate food, where climatic change, civil strife and other adversities inflict serious disruptions on food production and distribution, food aid may be essential to survival and avoidance of famine.

The UN World Food Programme [WFP] is the most efficient distributor of food aid, designed to provide relief or to support development. Food aid for relief alleviates acute stress among people whose food supplies are disrupted by war, civil unrest, enforced migration, severe climatic catastrophes, and/or massive crop failure. Food aid for development provides food as wages for economically productive work and to support basic social services. Food aid for development may support labour-intensive agricultural production, construction of roads, rural infrastructures, amenities and services essential to developing economies. Food aid helps to stabilise food prices, to provide for pregnant women, nursing mothers, school feeding and pre-school infants.

WFP administers food aid most efficiently. Misdirected, ill-conceived bilateral food aid can do more harm than good. One could cite a weary litany of bilateral food aid where indiscriminate dumping of donors' surplus crops and livestock products have depressed prices and discouraged agricultural production in recipient countries. Food aid administered by many bilateral agencies is motivated more by the donor's political and commercial interests than relieving acute or chronic distress among malnourished people. WFP resources have been seriously overstretched, because of exceptional demands resulting from military, civil and inter-ethnic strife and conflict, mass migrations such as in Darfur, cataclysmic disruptions as produced by the Tsunami. As extreme demands for food aid will inevitably persist, more generous and sustained support for the World Food Programme should be a high priority for all donor agencies.

What constitutes nutritional adequacy has attracted scientific and unscientific interest for many centuries. Alchemists searched for but never discovered the *Elixir vitae*. In Britain during the mid-nineteenth century desirable dietary standards were described as: "the least costly foods to prevent starvation and disease during massive unemployment". The League of Nations standards were intended "to marry health and agriculture". Our most primitive ancestors, by trial and error, identified certain foods as pleasant and digestible, others as obnoxious, insulting or injurious to health, foods acceptable and foods unacceptable. Our early antecedents chose their foods from plants and animals indigenous to their ecologies and climates resulting in widely different food preferences among nations and communities.

THE PHENOMENOLOGY OF FOOD SUSTAINABILITY

Sustainability is encountered with increasing frequency in a variety of arenas. As the global community of both developed and developing nations faces the challenges of finite natural resources, burgeoning public budget deficits, and growing interdependence and complexity, the need to do more with less has become the sine qua non of both government and private sector economic activity. Sustainability as an issue has both emerged from, and helped to shape, each of these challenges. For example, concern over the degradation and exhaustion of the environment has led to the explicit linking of economy and ecology, as necessary to sustaining growth while conserving the globe's natural endowments for future generations.

Tax-payer's "revolts" in the United States, combined with continued demand for public services, have forced government to reexamine the scope of its activities while seeking ever greater efficiencies. And in Eastern Europe, parallel but even deeper concerns have been expressed in city streets and squares, contributing to the dismantling of Communist dictatorships and their state-managed economies.

In the foreign assistance realm, the objective of making development investments more productive has taken on new urgency as developing country indebtedness reaches new heights, bilateral assistance declines, multilateral agencies try to divide their resources among ever more competing needs, and private capital and trade flows take on increased importance. Much of the task of developing in a sustainable fashion involves technical knowledge and skills, in the sense of specialized information and tools derived from the biological and physical sciences and applied to development and production problems.

In the agricultural sector, such things as biologically-based pest control, water-saving irrigation systems, crop diversification, or soils salinity management comes to mind. These often-termed "hard" technologies are indisputably important for growth, development, and sustainability in the sector. However, there is also a family of "soft" technologies, those deriving from the human and social sciences that play key roles. These tend to be overshadowed by their more visible brethren in the "hard" category.

The topic of this book is one of these "softer," less directly visible technologies, those relating to organizations and management. A key component of sustainable development is the institutional framework within which activities are conceived, planned, funded, implemented, and managed. While the technical quality and appropriateness of development activities are critical to success, sustainable institutions are equally so. To be able to grow and develop all economies need an infrastructure of institutions, both public and private, that operates efficiently and produces goods and services effectively. Institutions with these characteristics are particularly lacking in the Third World.

This book focuses on institutional sustainability and its role in agriculture and rural development. It concentrates on collaborations between international donor organizations and developing countries to design and implement projects aimed at introducing performance and capacity improvements, which in turn lead to sustainable institutions contributing to agricultural and rural development. Despite the drop in proportion of gross domestic product earned by agriculture relative to industry and services, many developing countries remain heavily dependent on agriculture. Especially for the bottom tier of developing nations, the vast majority of their populations pursue their livelihoods in rural areas, and their well-being depends on farming and allied occupations. Over the long run, maintaining and enhancing agricultural production demands a continuous process of acquisition, adaptation, and application of new technologies.

Technological change in agriculture has increased almost exponentially in complexity and speed, making the world marketplace for agricultural commodities increasingly volatile. In some cases new technologies have placed previously unique or distinct commodities in direct competition with either natural or synthetic substitutes. For example, the use of biotechnology to produce natural vanilla flavour in U.S. laboratories threatens a $50 million export for Madagascar, where 70,000 small farmers produced three quarters of the world's vanilla beans. In other cases, the failure to adopt a new knowledge has placed large populations at risk of malnutrition and even starvation.

This is most vividly true in sub-Saharan Africa, where institutions have not been able to support a "green revolution" of increased food production based on new varieties of food plants and improved agricultural techniques. Dynamic, capable institutions that deal with this increased complexity and rate of change on a sustained basis are central to successful agricultural development.

Recent studies in the United States, notably that of Belasco, have drawn attention to the way in which changes in the food industry have provoked resistance from some sections of consumers, and the demand that government intervene more to protect the interests of the food consumer. The food industry, in turn, has counter-attacked in a number of ways to the barrage of criticism and public consciousness surrounding food issues. Belasco details these 'counter-revolutionary' (or perhaps 'counter-counter-revolutionary' would be more accurate!) strategies: food manufacturers have sought to dilute recommendations ('reduce' rather than 'avoid' foods that are bad for you); they have employed what marketing people call 'repositioning', altering the way products are described rather than the products themselves; and they have increasingly employed strategies both to differentiate products and to differentiate markets.

In the United Kingdom today we could scarcely fail to be aware of the importance of product differentiation: the food industry has employed advertising techniques to differentiate between products for many years. However, what is new, and very much a response to consumer resistance, is product differentiation in 'healthy' or 'ecologically-friendly' foods, an important aspect of what has been dubbed 'Green consumerism'. Similarly, the extent to which the market can be differentiated according to the tastes of different groups of consumers, has received widespread attention, notably in the food trade journals. We would argue, in fact, that some segments of the market in developed countries have entered a 'post-fordist' era, as far as the consumption of food is concerned. That is, an era not so much of the mass consumption of homogeneous, industrially produced food as of the development of healthy foods for market 'niches', often reflecting ethnic variety and traditions, but utilizing the full armoury of the food processing industry, and targeted to consumers willing to pay for high value-added products. With the growing sophistication in manipulating the chemical constituents of food, highly processed food can be presented as 'healthy' food, manufactured to have low fat, low cholesterol and high fibre content, in line with current dietary recommendations.

The challenge facing the food industry, and this is linked to shifts in the food system more generally, is how to make healthy foods more marketable, and marketable foods appear more healthy. Both questions are capable of eliciting a response from food manufacturers, but to understand the background to this response we need to examine the patterns of resistance which have grown up to recent changes in the food industry, patterns which echo some of the forms of resistance to environmental changes.

The beginnings of what Belasco calls a 'countercuisine' are linked to a number of quite important social changes which, as we have already noted, shift the individual's attention from the sphere of production relations to those of consumption. Three aspects require our attention: the way in which we arrive at our impressions of what foods to avoid and what foods to consume (the quality of food as a commodity in the market); the way in which food 'signifies' and expresses other aspects of our lives (food as self-enhancement); and finally, the way in which our consumption of food is linked to broader questions of ownership and organization in the economy as a whole (food as part of a broader political economy).

Almost unnoticed, issues surrounding each of these aspects of the consumption of food have assumed enormous importance in recent years, but close examination reveals that what at first appear to be almost random evidence of social resistance to the modern food system have their roots in fundamental social changes. These social changes need to be outlined, and their implications examined. It is our contention that too much attention has been devoted to the 'signifying' aspects of food and diet, the semiology of food consumption, by sociologists and anthropologists, to the detriment of a wider understanding of the social transformations implied by the economic and technological changes in the food system.

As we have seen, women's roles have been transformed in a number of ways, some of which are clearly irreversible. Domestic work, including housework, has been opened up to important market forces, and the manufacture and purchase of 'white goods' for the home is now an important activity. At the same time the idea that housework is no longer onerous has served the interests of manufacturers and others. It is far from clear that technology in the home is labour-saving; this assumption, needs to be examined critically. Evidence exists that a human price is exacted for the benefits of speed and convenience which 'white goods' bring. Housework standards change, and women expect (and are expected) to raise the standards of housework in turn. At the same time, it is likely that women's release from the worst forms of labour drudgery is associated with more time being spent by them on other activities in the home, particularly attention to children.

This is not to say that many household tasks are not easier; they are. It is merely to point out that the definition of housework, and with it women's responsibilities in the home, is constantly shifting. It is not immutable. It is also clear that since the 1970s important changes have occurred in popular understanding of the relationship between food and health. Preventative medicine may only be in its infancy, and receive little official encouragement, but for some groups of people in the industrialized world, healthy eating is now considered essential. Healthy eating in the past depended critically on local custom and diet and, most importantly, income. Today this is still true, but local variations have less importance, and additional factors have made their appearance, which have engaged the attention of a battery of professionals, such as health educators, doctors, consumer groups and alternative therapists. The perceived need to diet to reduce obesity has given rise to a huge consumer market in low-calorie foods, including the controversial 'low-cal' liquid diets and innumerable diet 'systems'.

The publication of evidence about heart disease and cancer has led to some marked shifts in the public consumption of certain foods, especially in the United States, where the changes date back to the publication of the McGovern Report in 1977. Food has become one of the obsessions of our time, associated as it is with good health, longevity and the occurrence of stress. Physical fitness has also become an obsession in some quarters, perhaps also a 'fad'. Evidence from the United States suggests that, while in 1960 only a quarter of adults exercised regularly, by 1980 over half did so.

The point is not that everybody is exercising more, but that more people are, and that health is increasingly linked to diet in many people's minds. This is not a trend likely to be discouraged by the food industry, since it also represents a change with commercial possibilities. Indeed, the giant food firms have jumped on this bandwagon, adapting to each new nutritional recommendation—low sodium, low saturated fats, high fibre—to the point of confusing consumers grappling with labels and elusive media-speak, such as 'lite' and 'natural'.

Changes have also occurred in public perceptions of the two processes: appropriation and substitution. Following the scare over DDT between 1969 and 1972 in the United States, and subsequent publicity about the use of other pesticides, attention has been given to the effects of pesticide residues in food, especially by the London Food Commission in the United Kingdom. Few studies have been undertaken, and even fewer given publicity, about the relative nutritional quality of organically produced and chemically produced food. Pesticide use in the United States rose 500 per cent between 1950 and 1986, but in the latter year one-fifth of United States's crops were lost to pests, the *same percentage as in* 1950! The tactic employed by the food industry in meeting the criticism of environmentalists and whole food enthusiasts, has been that the industry can 'compensate' through industrial processes for the shortcomings of nature.

The industry argued that organic foods were more expensive (ignoring the hidden subsidies provided to chemically produced food), that organic foods could not meet market demand, that food additives increased the palatability of food, and that convenience removed the drudgery from food preparation. By the late 1960s in North America, and perhaps a decade later in the United Kingdom, food manufacturers began to 'reposition' themselves, as well as their food products. Now they were on the side of emancipated women and environmentalists; they offered themselves as accomplices in women's drive for more independence and in the vanguard of responsible, sustainable resource management. Market researchers have credited much of the improvement in the food industry to their own responsiveness to consumer pressure. In the words of an industry spokesperson:

For example, ten years ago the only place people could find additive-free *natural* products were in the comparatively cramped and small premises of health food stores. Today such products, not only much improved and—dare we say it?—engineered to satisfy consumer preferences, are mainstream products in prime placing in major supermarket chains the length and breadth of Britain.

This represents one side of the story, but in practice healthier eating has also presented a challenge to the food industry which it has found difficult to grasp. The problem, in a nutshell, is how to leave out food additives and processing, and still add value to the final product.

The post-war period had seen additives 'substitute' for natural ingredients, while much that was natural had been removed from food. Food was given more value added through packaging, processing and ensuring it stayed 'fresh' longer. The challenge represented by making 'healthy' foods from 'health' foods was how to make profits by appearing to do less to the product. The industry responded to the challenge, as we have noted, by both market and product differentiation—identifying, and exploiting market opportunities for 'niche' products, such as low-calorie foods, ethnic health foods and fitness-related foods.

Finally, the food industry has been in the forefront of the promotion of food supplements; people can now supplement their inadequate diet from a range of commercially promoted food 'accessories'. In 1985 over US$3 billion was spent in the United States on vitamin and mineral supplements and over $4.5 billion in supplements to fortified breakfast cereals. This diet supplementation was the inevitable outcome of the way eating had changed in the United States. In 1909, 40 per cent of calories had been provided from fruit, vegetables and grains. By 1976 only 20 per cent of calories came from these sources, the rest from fats and refined sugars. As people ate more calories they ate fewer vitamins and minerals—supplements have helped to fill the vacuum left by changes in the 'affluent diet' of North America.

Another important social change that has accompanied the shifts in diet referred to above is that governments in the industrialized countries have been forced to make some accommodation to the pressures put upon them by food consumers, and interest groups representing consumers. In the United Kingdom this has come about largely as a result of successive food 'scares', such as those over salmonella, listeria and bovine spongiform encephalopathy ('mad cow disease') in the last few years. The extent to which governments will respond to consumer anxieties is still unclear, but the current debate about public nutrition policy and food legislation represents an excellent illustration of the kind of contradiction to emerge from the development of the modern food system.

The demand for more intervention in the food system presents difficulties for governments committed to the idea of choice in the marketplace. On the one hand it is difficult for them to resist, as a matter of principle, the demand for more information about food products. Issuing 'guidelines' about food and health is not inconsistent with conservative ideology, which places individual choice before government controls. The publicity given to the McGovern Report in the United States after 1977 served to divide the food lobby. Part of the food industry could take comfort from the encouragement being given to people to eat more cereals, vegetables and fruit. Another section of the industry tried to develop marketing strategies to overcome opposition to increased consumption of other items (particularly meat, dairy products and eggs).

The development of a 'fordist' diet in the 1950s and 1960s had meant that red meat consumption rose, in the United States from 121 Ibs per capita in 1930 to 160 Ibs per capita in 1970. By 1968 20 million tons of vegetable protein was being fed to livestock to produce 2 million tons of animal protein. Since 1980 there has been a steady reversal of this trend: beef consumption has fallen in the United States, and voluntary reductions in food consumption, like those recommended by the McGovern Commission for eggs, sugar and animal fats, have also begun to be registered. However, this, too, is only part of the total picture.

Changes in diet, encouraged by public debate, have not yet pushed or persuaded governments, in the United States or Western Europe, to take a more interventionist stand on other aspects of food consumption, notably the use of food additives, although recent legislation like that proposed in the United Kingdom in March 1990 would introduce more rigorous standards of hygiene in the food industry. In the United States the Federal Food and Drug Administration (FDA) co-operates very closely with the food industry. In practice many of its recommendations are welcomed by the industry because they reinforce the idea that most food already on the supermarket shelves is fit for public consumption.

The FDA-initiated testing of food led to the conclusion that over a thousand substances used as food additives were 'generally recognized as safe' (GRAS) for use in food, whereas only two substances were banned as a result of FDA testing: a carcinogenic weed killer and, in 1969, cyclamates. The United States Government has proved reluctant to act as a watchdog on food additives. Meanwhile over 1 billion Ibs of additives are added to food products in the United States, equivalent to 5 Ibs of additives per capita. It is important to consider another aspect of food consumption: the meaning people attach to it, and the way in which food consumption practices help structure social relations more generally. Gofton has made a useful first attempt to come to grips with some aspects of the sociology of food consumption in the United Kingdom today. He notes several of the trends referred to above and in other chapters of this book: the growing concern for health; home-centred living; redefinitions of social roles within households; the trade-off between the time taken in food preparation and leisure activities; and the emphasis given in our society to individualism and hedonism.

Many of these changes are linked to the way society has become organized in the developed countries, such as the shift towards a more informational society where the production of information rather than goods is at a premium. Similarly, there has also been a shift towards relatively decentralized authority systems, particularly in the family, where individual members are more likely to make autonomous decisions about (among other things) what and when they eat. Finally, during the 1980s we witnessed in the United Kingdom what has been dubbed the beginnings of 'the enterprise culture', which placed importance, particularly in ideological terms, on individual entrepreneurship and initiative rather than bureaucratic surveillance and government responsibility. Gofton argues, convincingly we believe, that patterns of food consumption (and, we should add, patterns of resistance to highly processed, 'industrialized' foods) are closely related to changes in social behaviour and organization, which place private consumer choice and individualism before community-based or class-based patterns of behaviour. The trends in consumption, implied by different market 'niches', together with underlying shifts in the way households and 'communities' are constituted around specific interests, are important components of post-industrial society.

Gofton also reviews various strands of what we might term the 'sociology of food' focusing on three major areas in which attention to food might be linked to 'post-industrial' and 'post-modernist' concerns: the semiotics of food (represented by Mary Douglas in particular); the debate initiated by Pahl and Wallace and Gershuny about changes in the way the household is constituted; and the work of Bourdieu on the relationship between upbringing, class and taste (including taste in food). Something of this approach is reflected, within British anthropology, in the work of Jack Goody.

The work of Pahl and Gershuny, important as it is, is principally concerned with work rather than food, but the other sociological currents certainly reflect serious attention to food consumption habits, and the role of these habits in our culture. These approaches explore the phenomenology of food consumption, the way in which food tastes and habits operate not simply as indicators of lifestyle but also as ways of conveying meaning to experience. Other contributors to this debate include Charles and Kerr, whose recent work has addressed the relationship between eating habits, gender relations and the class structure, basing their analysis on empirical evidence drawn from interviews with over 200 women in the United Kingdom between 1982 and 1983. Charles and Kerr take as their point of departure the way in which food practices contribute to the reproduction of the social order, drawing on Bourdieu and Althusser in stressing the role of ideology in social reproduction. Their analysis places considerable emphasis on the meanings carried by food about the nature of the social occasion, and the way these meanings provide the ground rules for social structure, gender roles and class identities. This kind of research is phenomenological in outlook but essentially empirical, even occasionally empiricist, in design. It is grounded less in what Giddens has termed the 'knowledge ability of [the] agents' themselves and more in the unconscious behaviour of these agents, whose cultural patterns help to define them within their society.

Research such as that undertaken by Charles and Kerr delineates some important aspects of social behaviour in Britain today, but we would argue that it fails to relate behaviour within the family to wider structural changes in the role of food within society, and to the expansion of the food system. Reading Charles and Kerr one could be excused for thinking that working-class people never ate 'fast foods' (there is no entry on fast foods in the index) while 'convenience foods' are relegated to just three pages of text. Apparently the women interviewed used convenience foods, but not in the main meal they prepared—convenience foods were not 'proper meals'.

Expending time and effort on the production of a meal in some way conferred goodness on it: perhaps a moral rather than a nutritional goodness but this distinction was not made by the women...A meal consisting of convenience foods...cannot constitute a proper meal, is by definition less healthy than a proper meal and involves much less preparation time...We may therefore see women's concern that the family should eat properly and healthily in terms of home cooked food provided at regular intervals as having been handed down through family tradition; it is a central part of family life. This kind of analysis places evidence of continuity before evidence of change, and imbue the meanings people attach to their behaviour with more importance than the behaviour itself. It is also instructive to note how the authors have renegotiated these meanings in the text.

It is not easy to establish from their work whether women are making more use of convenience foods, although the suggestion is that they probably are, because the analysis is largely concerned to demonstrate that any resort to convenience foods needs to be located within traditional family behaviour. Emphasizing that people look upon food as 'natural' should not blind us to the fact that in practice it frequently is highly processed. What people *feel about food* may be important, but so is the evidence about *what they eat.* One suspects that if the same women were talking to researchers less concerned to establish its meaning for their argument about gender and class relationships, it would have been much more difficult to discuss food preferences in terms of the provision of a 'proper meal'.

Instead of evidence for the reproduction of the existing social order, we would see evidence of change, in the social changes surrounding the way food is consumed as well as in the food itself. Phenomenological approaches to the meaning of food consumption would benefit from taking some cognizance of the processes through which food itself has become an industrially produced good.

This book has examined the effects of the modern food system in accelerating social changes, in both developed and developing societies. We have argued that the ideological aspects of the food system are not confined to those usually reported on by sociologists: the 'appropriateness' of food, and the meaning attached to food in everyday life, within the experience of working-class people. It is our contention that changes in the material basis of food production and consumption are linked to the transformation of 'nature', and that the establishment of new areas of commodity production influences the values attached to social relations.

Capitalism is an evolving system, whose values and practices do not lie embedded in the past, anymore than the behaviour of working-class people is lodged in static definitions of what is an appropriate, definition formed by their parents and grandparents within a quite different social milieu. Food has become fully commoditized; it is therefore to be expected that it's 'naturalness' would be emphasized by the food industry, in the process generating new symbolic meanings as a path to further accumulation.

As Massey has observed there needs to be a debate about the 'natural', which can draw on recent post-structuralist thinking. The 'denaturalization' of space, in the form of attention to 'location theory', was a preoccupation of some social scientists, particularly geographers in the 1960s. What is required now is similar attention to the way in which naturalization accompanies changes in technology and social relations. Nature has been refashioned in several ways, which this book has drawn attention to within an integrated framework of analysis.

1. Within the labour process the employment of women has, through routinization, become naturalized as women's labour has shifted into and out of the home.

2. Technological changes in agri-food production, now enhanced by biotechnology and genetic engineering, have manipulated biological processes, appropriating some and substituting others, redefining and refashioning nature as a source of profit and capital accumulation. As the food we consume has become more processed it has been presented as more 'natural' by the food industry.
3. Finally, the development of modern agriculture has forced a separation between the 'countryside' as an area of consumption (amenity, recreation, aesthetics) and the production of food and fibre from the land. The drive to protect the countryside, or more generally to seek greater protection for conservation and wilderness areas in other industrial societies, is defended and advanced on the basis of conserving what is 'natural', and a true embodiment or representation of 'nature'.

It is clear from this analysis that as the modern food system becomes more integrated, and the management of contradictions, such as those presented by the environment, become more urgent, the more the food system departs—in all three senses referred to above—from being 'natural'. Naturalization is the price society pays for seeking to transform its image of itself. Or, as Bourdieu has expressed it, each system produces the naturalization of its own arbitrariness. We have argued that many of the contradictions posed by the modern food system can be handled by policy interventions, such as those to contain environmental costs and to address public concern over the quality of industrially produced food. The modern food system developed out of a combination of factors. These included: state intervention in agricultural policy designed to stimulate the adoption of new technology; guaranteed markets; the food processing industries' access to female labour, partially released from full-time employment in the home by (initially) wartime restructuring; and changes in family farming which served to legitimate the demands of more 'efficient', small-scale but highly capitalized producers. The management of the contradictions of the modern food system can be handled in the industrialized countries of the North, at a price, for several reasons that require close attention.

As we have seen, productivity increases in agriculture have been subsidized through industrial accumulation in the countries of the North. In the post-war period of intensive accumulation, the significance of cheap food arises not simply, as initially understood from its role, as a wage good; but also from its contribution to a widening of the market for consumer goods in general, which has accompanied the growth of labour productivity within the advanced capitalist societies.

The costs of this policy, of continuing to support agricultural modernization, have been passed on to the consumer and taxpayer in the European Community, the United States and Japan. It is increasingly clear that this agricultural policy model, even when only partially understood, is being called into question, notably by conservative critics as well as those with 'Greener' credentials. At the level of ideological legitimation, which has remained important throughout the post-war period in helping to define the relationship between farmers and the state, the resistance of public opinion to the priority attached to this model, at great environmental and social cost, has become an increasingly important part of the new political agenda of the 1990s.

Furthermore, the rise to importance of what we have termed 'environmental managerialism', in the industrialized countries, represents a significant policy opening, which could not have been anticipated a decade ago. The improved management of 'externalities' associated with the food system can reduce the full impact of the transformation of the countryside, through negotiated accommodations to environmental concerns. In the United Kingdom we have been conditioned to accept a managed countryside for a least a century; in other countries of the European Community the emphasis on preserving the cultural association between family farmers and the environment is equally strong politically.

Indeed this nexus of interests recently gained prominence as the philosophical foundation of the EuropeanCommi-ssion's proposed rural development policy, as outlined in *The Future of Rural Society*. In North America, the issues are gather different, in that the separation of wild lands (to which public access is considered important) and agriculture, is more pronounced. Nevertheless, here too the armoury of interventionist policies designed to manage the contradictions of development is receiving more attention: low input farming, wilderness retention, the establishment of recreational priorities for an affluent, urban population.

Finally, as we have seen, in the industrialized countries of the North there has been considerable resistance to the complete industrialization of the diet. For a substantial minority of the population—in the United States usually considered to be about a third of the market—'healthy' eating has become an alternative to the excesses of the 'fordist' diet. The food industry has accommodated, indeed sought to exploit, many of these changes, whose origin we see in the wider educational opportunities of the post-war generation, as well as greater private affluence. As we have seen, it is possible to increase value added in food products without necessarily losing the nutritional value of food, although this poses real problems for the large food conglomerates in increasingly competitive, oligopolistic markets.

Within some sectors of the durable consumer goods industries the development of 'healthy' convenience foods can also provide a stimulus to production, microwaves and toasted sandwich-makers replace or supplement other white goods, but without leading to the consumption of more unhealthy food. Changes in diet have also been prompted by wider changes in public attitudes: the increases in vegetarianism, the concern with animal welfare, the widespread interest in high-fibre, low-calorie foods as part of attempts to lose weight. These changes are significant, and are most unlikely to be reversed: they have helped to set in motion a series of effects on the supply side of the food system, which are reverberating and will continue to reverberate during the remaining years of this century and into the next.

Within the developed countries of the North, then, the modern food system poses problems for the economy and society which call for changes in policy, and the management of the environment, but which can be successfully addressed, albeit at a price in both financial and political terms. The implications of the food system for the South, we would contend, are altogether more radical. It is worth considering the differences carefully.

In the South the modern food system can only be partially adopted, because structural inequalities have prevented the industrial sector from acting as the 'motor' for the restructuring of agriculture around industrially produced food for mass consumption. In most developing countries the agricultural sector, and particularly staple production, has been appallingly neglected, not least in terms of the impact of agricultural intensification on the poor and the environment.

Where rising agricultural productivity is supported by state subsidies, or subsidies from the developed countries and their institutions, it has benefited some groups much more than others. The 'fordist' diet, exemplified by the penetration of the grains-livestock complex, has only reached a very narrow band of food consumers in even the so-called 'middle-income' countries of Latin America. In most of Africa and Asia, the fordist diet remains part of the mystique of Western cultural values, as distant from its audience as the movie star.

Environmental managerialism, too, does not represent a workable strategy for most governments in the South, where it has neither tradition nor institutional foundation. In the North, the improved management of the environment, usually at the behest of relatively affluent, urban groups, is endorsed with relatively little political dissent; the modest interventions to ensure a better management of the countryside, or conservation of wilderness, have overwhelming public support in Europe and North America. In the South, on the other hand, land degradation is intimately linked to poverty in rural areas. A vast, largely rural population of poor people, many of them *very* poor, looks to the environment as the basis of its precarious livelihood. Environmental problems in developing countries can only be realistically understood as the outcome of underdevelopment and poverty. The 'environment' lacks ideological resonance; it is to be 'tamed' or 'conquered', and made to yield economic revenues and livelihoods. Environmental managerialism, without its ideological underpinnings which explain its strong cultural appeal in the North, would need to be radically modified to make any impact on the causes of environmental degradation in the South.

It can be argued, indeed, that attempts to manage the environment in developing countries, in as much as they are predicated on Northern experience, are not merely unworkable, they are also frequently prejudicial to the interests of the poor. In the Brazilian Amazon the disputed land and natural resources cannot be 'mediated' successfully by international agencies alone, whose principal concern is thought by many Brazilians (on both left and right of the political spectrum) to be the protection of the quality of life in the North, rather than the rights to livelihood of poor people in the South.

Many Brazilians say, and their views are at least partly convincing, that campaigns to protect the Amazon rain forest are the new face of 'neo-imperialism', prompted by fears of global warming (largely exaggerated in view of the much more serious effect of fossil fuel emissions) and the perceived need to preserve rare animal and plant species, rather than concern about poor frontier settlers or the pernicious social impacts of the Brazilian debt.

In many parts of the South, 'resistance' is to be understood primarily in terms of the right to livelihood, to citizenship and to the recognition of basic human rights. Struggles such as those surrounding land invasions, or the protection of indigenous people and rubber tappers in the forests, represent struggles for existence, rather than movements to secure improvements in the quality of life. The development of the modern food system is linked to these struggles, albeit sometimes in tenuous ways, but the management of the environment cannot resolve the contradictions implied by the development process. Indeed, the effects of recent structural adjustment policy, and attempts to manage the enormity of the debt burden in Latin America and Africa in particular, have intensified environmental problems, and make any 'solution' that does not include major changes in these policies largely irrelevant. The implications of the partial adoption of elements of the modern food system in the South can only be understood, and the resulting problems successfully addressed, by changes in structural policies encompassing the global economy. They cannot be resolved, as to a large extent they can in the North, where they occur.

Finally, as far as changes in the diet are concerned, the partial development of the modern food system in the countries of the South has served to replace (and in some cases destroy) existing, indigenous food systems. This has occurred without effective resistance from a broadly based mass movement of consumers, educated as to their rights, and able to exert pressure on government to accede to their demands. It is virtually impossible to envisage public campaigns over salmonella or food additives in Mexico or Indonesia, or, what is altogether more telling, effective mass movements to introduce new standards of food hygiene in shops or on street corners. Meanwhile, food irradiation—increasingly contested by the radical food lobby in the North—opens up market opportunities in the South.

It is clear, then, that what is at issue if we consider North-South relations, and examine the implications of the development of the modern food system within an international framework, is a set of concerns which stretch the parameters of our model, and our analysis. To speak of resistance and struggle in the South, is to speak of the survival of alternative systems which link the production of food directly to its transformation and consumption, systems which embody quite different patterns of social reproduction, systems which—in the form of ubiquitous 'externalities'—carry many of the environmental and social costs of our 'successful' development model.

Our ability to grasp the fact that the economic and political hegemony of our values, and our economic policies, have served to undermine the integrity of other food systems, remains the major challenge we in the North face today. For the spread of the modern food system marks the success of the North's cultural hegemony, as well as its economic supremacy. We need to address the sustainability of our own models, then, before we are in a position to hand them down to others.

ESTIMATES OF HUMAN NUTRIENT REQUIREMENTS

Estimates of human nutrient requirements were first codified during the latter half of the nineteenth century. From the time of their establishment, FAO and WHO have sponsored many meetings, committees and working groups to study and recommend desirable nutrient intakes, each meeting re-estimating recommended levels based on new knowledge. At least 50 national, regional and international agencies recommend what they consider desirable intakes of essential nutrients. Recommendations are customarily cited as specific amounts of each essential nutrient to be ingested daily. Some authorities name them Recommended Daily Allowances [RDAs], others Recommended Daily or Dietary Intakes [RDIs], others Recommended Nutrient Intakes [RNIs]. The evolution of recommended dietary standards was reviewed by several authors. The purpose of RDAs is to encourage healthy eating habits and to prevent diseases attributable to nutrient deficiencies. Earliest motivations for government dietary standards arose in times of severe stress and distress; to feed nations at war or when suffering economic depressions, widespread poverty and malnutrition.

Recommended Daily Allowances [RDAs] the preferred American designation, are intended to serve several purposes: (a) to assess adequacy of food intake among selected population groups; (b) for national food and agricultural policy formulation; (c) to balance food nutritional quality with human need; (d) to provide adequate food during emergencies, particularly for the poor and disadvantaged; (e) to prescribe and enact standards for nutrient labelling of packaged foods. RDAs were first designed to meet the needs of the majority, often embodying margins of safety to allow for variable needs among people of different ages, activities and conditions of health. It is now customary to prescribe specific RDAs for citizens of different sexes, ages and levels of activity.

RDAs were first calculated by estimating the average physiological requirements for each absorbed nutrient, a value then modified to compensate for incomplete absorption and utilisation, variability among persons, bioavailability among different food sources. RDAs are based on foods consumed in normal diets, not on pure substances or concentrates, and take account of nutrient variability within and among particular foods. At high levels of intake some nutrients, such as vitamin A, may be toxic so that upper limits of safe intake are prescribed.

It is difficult to prescribe precise RDAs for all conditions of men, women and children and their variable activities. Experiments on human beings are expensive, time consuming and only small groups of volunteer subjects can be studied in confined controlled conditions. Using humans as experimental animals also raises ethical objections. RDAs are intended for people as they go about their normal daily activities but food intakes and nutrient balance are difficult to determine where subjects move freely and are not confined in a metabolic research facility.

Data based on *ex post facto* recall by subjects going about their normal lives are at best approximate. Despite a vast accumulation of nutritional data, with more sensitive analytical tools available, nutrition remains a relatively imprecise and speculative science.

Concepts change as more evidence comes to light. Thirty years ago it was believed military men working in Arctic climates needed more food than those working in temperate zones so Canadian troops in the Arctic north were given packaged rations that provided 5,000 kg Cals/day. It was later realised that with insulated clothing and the need to move slowly to avoid heavy perspiration that could freeze, caloric requirements were no higher in cold than in temperate climates. Since caloric intake must balance energy expenditure, people who perform heavy work or strenuous exercise need higher caloric intakes than sedentary persons.

People who change from energy-intensive activities to a sedentary life-style tend to eat more than they need. Overeating with resultant obesity is disturbingly evident among the richer nations and affluent persons in less developed economies. It is reported that between 25 and 30 per cent of Europeans and North Americans are excessively obese, at risk of cardiovascular diseases, diabetes and other related disabilities.

The obese among the affluent reflect Nerissa's comment in Shakespeare's *Merchant of Venice*: "They are as sick who surfeit with too much as those who starve with nothing."

- Variability in Composition: The nutrient content of every food crop varies among genotypes and conditions of cultivation. Among the world's wheat collection on a dry weight basis, protein contents range from 6 to 23 per cent, lysine (as per cent of protein) from 2 per cent to 4.3 per cent. A review of over 1,300 sorghum analyses reported protein contents (N × 5.7 DWB) to range from 7.1 to 14.2 per cent, lysine (mg/g nitrogen) from 71 to 212, iron (mg/100g) from 0.9 to 20.0, niacin [mg/100g] from 2.9 to 6.4. Significant variability was reported among samples of pearl millet (*Pennisetum typhoides*), chickpea (*Cicer arietinum*), faba bean (*Vicia faba*), and lentils (*Lens esculenta*).Accurate chemical and biochemical analyses are difficult and demand professional competence and experience. Inexperienced technicians are unaware of many sources of error among standard analytical methods. Some 16 sources of error were identified in the classical Kjeldahl method of determining protein nitrogen.
- Nutritional Fashions: Responsible nutritional scientists seek to encourage rational eating habits based on the best available knowledge. The alchemist's pursuit of the *Elixir vitae* is far from moribund among publicists who proclaim all manner of diets guaranteed to ensure good health and longevity. The penchant for nutritional panaceas has been evident for the past 50 years. During the 1950s vitamins were in favour, in the 1960s proteins and amino acids were in fashion, from the 1970s cholesterol and cardiovascular dysfunctions obsessed public attention, next came the health benefits promised by dietary fibre. The present passion is for nutriceuticals, proclaimed in principle some 4,000 years ago by the Chinese, and for 'functional foods' (which beg the question: what are non-functional foods?), neither of which are specifically defined.
- Recommended Daily Allowances (RDAs): RDAs, RDIs and RNIs can be useful as general guides for cognisant officials responsible for formulating national health, food and agricultural policies. For most consumers, RDAs have little practical meaning. Consumers are better advised by illustrated guides to healthy eating based upon mixed diets of foods that are familiar, acceptable and seasonably available. Human nutritional requirements, criteria used to arrive at recommended levels of intake, consequences of serious deficiencies, food intolerances and allergic responses are discussed in greater detail in other publications.

- Urban Growth and Food Security: Predictions of the size to which the world population will grow, and for how many persons the planet can provide sustainable food security 50 or 100 years from now, though intellectually diverting, are of little practical utility. One can estimate with reasonable certainty what could be produced with present resources. What may be feasible 50 or 100 years from now is indeterminable.Of more pressing urgency are the consequences of rapidly changing demographics and rising disposable incomes among urban communities. More than half the people of Asia and Latin America live in large towns and cities, expanding in size and populations all over the world. Though some food can be raised in city gardens and horticultural allotments, urban communities rely on food safely transported from rural farms, inland and maritime fisheries.As people migrate from rural to urban communities and disposable incomes rise, eating habits change. Growing demand for processed convenience foods is illustrated by the replacement of cereals, pulses and cassava, hand ground and cooked into some form of porridge, with processed foods such as bread, biscuits, cakes and pasta. Of critical consequence for natural resource conservation is the growing demand for livestock products, which across SE Asia is rising by over 20 per cent a year.Few of the world's cities were planned and laid out in expectation of the volume and density of vehicular traffic destined to grow progressively worse in all cities on all continents. To alleviate traffic congestion, the City of London Authority has imposed heavy tolls on private vehicles that enter the city on working days. Too little investment is devoted to improving and maintaining rural infrastructures, to rural-to-urban and intra-urban transportation systems. The longer time taken to transport perishable food materials from rural producers to urban processors and markets, the greater is the probability of biological and economic loss.Urban food security demands consistent access to a safe adequate supply of acceptable, affordable foods dependent on a. technologically efficient systems of preservation and processing, primary preservation being applied soon after harvest; b. logistically efficient and economic means of storage and transportation; c. reliable markets which retail foods that are safe, of acceptable quality, in unit quantities and prices that all segments of urban society can afford; d. national food and agricultural policies that provide and maintain the infrastructures, services, transportation facilities, social and economic conditions essential to efficient holistically integrated food systems.

SUSTAINABLE AGRICULTURE, FOOD SECURITY AND CLIMATE CHANGE

The agricultural sector is a driving force in the gas emissions and land use effects thought to cause climate change. In addition to being a significant user of land and consumer of fossil fuel, agriculture contributes directly to greenhouse gas emissions through practices such as rice production and the raising of livestock; according to the Intergovernmental Panel on Climate Change, the three main causes of the increase in greenhouse gases observed over the past 250 years have been fossil fuels, land use, and agriculture.

LAND USE

Agriculture contributes to greenhouse gas increases through land use in four main ways:

- CO_2 releases linked to deforestation
- Methane releases from rice cultivation
- Methane releases from enteric fermentation in cattle
- Nitrous oxide releases from fertilizer application.

Together, these agricultural processes comprise 54% of methane emissions, roughly 80% of nitrous oxide emissions, and virtually all carbon dioxide emissions tied to land use.

The planet's major changes to land cover since 1750 have resulted from deforestation in temperate regions: when forests and woodlands are cleared to make room for fields and pastures, the albedo of the affected area increases, which can result in either warming or cooling effects, depending on local conditions.

Deforestation also affects regional carbon reuptake, which can result in increased concentrations of CO_2, the dominant greenhouse gas. Land-clearing methods such as slash and burn compound these effects by burning biomatter, which directly releases greenhouse gases and particulate matter such as soot into the air. Livestock and livestock-related activities such as deforestation and increasingly fuel-intensive farming practices are responsible for over 18% of human-made greenhouse gas emissions, including:

- 9% of global carbon dioxide emissions
- 35-40% of global methane emissions (chiefly due to enteric fermentation and manure)
- 64% of global nitrous oxide emissions (chiefly due to fertilizer use.)

Livestock activities also contribute disproportionately to land-use effects, since crops such as corn and alfalfa are cultivated in order to feed the animals.

Worldwide, livestock production occupies 70% of all land used for agriculture, or 30% of the land surface of the Earth.

AGRICULTURE AND FOOD SECURITY

Pre-Independence India suffered repeated famines, drought and food shortages. But following the Green Revolution in the '60s, yields and foodstocks rose manifold. Now, 30 years later, Indian farmers have realised the follies of their tryst with intensive agriculture. Despite 70 per cent of the population being engaged in agriculture and allied activities, declining foodgrain production and access to food remain the two biggest problems confronting the country. Liberalisation has made things worse: commercial crops are eating into the fertile land tracts meant for essential foodgrains. And six years after the World Trade Organisation came into existence, the anticipated gains for India from the trade liberalisation process in agriculture are practically zero.

Agriculture Minister C Subramaniam sat huddled with his senior staff. There was tension in the air. Depending on imported foodgrains coming from 12,000 miles away, particularly when it came in driblets following US President Lyndon Johnson's directive to "teach India a lesson", was humiliating enough. What worried him was that food stocks had reached such a precarious low that they were only sufficient for another two weeks. Still worse, there was nothing in transit. Recalls Subramaniam, "As a last resort, I told my officials and experts to identify the nearest food-carrying ships on the ocean throughout the world. I said we would identify those carrying wheat to other countries and appeal to the US President to divert them to India if other countries could wait for another six to eight weeks."

This was in 1966-67, the critical year of drought when India imported 11 million tonnes of foodgrain. A year earlier, India had imported 10 million tonnes of foodgrain. Around this time, the Paddock brothers, often referred to as 'prophets of doom', concluded in their book, Famine 1975, that by the mid-'70s at least half of India would be led to a slaughter house. Considering that India had a long history of hunger and famine, and the subsequent failure of the government's effort to boost agricultural production in the post-Independence era, not many disagreed with the conclusive analysis drawn by William and Paul Paddock.

After all, for a country literally surviving on a 'ship-to-mouth' existence, there was little hope. Soon after Independence, India had begun to seek food aid and in fact emerged as the biggest food importer of this century. In 1951, India got ad hoc assistance of 2 million tonnes of foodgrain to tide over the crisis arising from severe drought in several parts of the country. This was the first major food security on a significant scale after Independence.

Before 1947, Indian history was replete with famine, drought and food shortages. Between 1770 and 1880, as many as 27 food scarcities and famines were recorded. At least 20 million lives were lost in India in about 20 famines that had struck since 1850.

Much of this loss was because of misplaced colonial policies which aimed to derive maximum economic gain at the cost of human suffering and misery. After the British had created a transport infrastructure in the first half of the 19th century, they began encouraging farmers to grow crops that could be exported. The boom in export trade accompanied by rising prices made farmers shift to cash crops like cotton, indigo, poppy and sugarcane. The area under foodgrain subsequently shrank. In other words, efforts to improve agriculture in colonial India were directly linked to the needs of British industries.

In the first half of the 20th century, India's foreign trade showed the same colonial pattern with exports mainly comprising foodgrain and cotton, jute, oilseeds, opium and indigo. Against this, imports consisted mainly of consumer goods. Meanwhile, the cultivator's preference for commercial crops continued.

As is happening now, the trend during the pre-Independence years showed that the best lands available were used for the cultivation of commercial crops. The cultivation of foodgrains was relegated, to some extent, to inferior lands, the yield of which is much less even in normal years. The result was a steady decline in foodgrain output till the Second World War. Per capita food availability from domestic production too had declined. As B M Bhatia observes in The Famines of India: "India kept the precarious balance between increasing population and declining food production partly by turning from an exporter into a net importer of foodgrains but largely by the easy, though dangerous, expedient of pushing down the already low per capita cereal consumption to still lower levels."

The big gap between minimum requirement and supply continued throughout the post-War period. Till the 1960s, Indian agriculture was not sufficiently able to meet domestic requirements and the country had to rely on foodgrain imports.

SOWING THE SEEDS OF THE GREEN REVOLUTION

Realising the disastrous consequences of the growing gap between the rates of growth of population and food production, a vigorous 'grow more food' campaign was launched. The key goal of economic planning shifted to attaining self-sufficiency. The First Five-Year Plan, following food imports in 1951, gave maximum importance to the growth of agriculture. Among the measures initiated in the '50s to stimulate foodgrain production were land reforms, irrigation, fertiliser production, strengthening of research and the organisation of a national farm extension service. Consequently, increased harvests in the mid-1950s relieved the economy.

It was around this time that the government adopted the US Land Grant model of agricultural research. Agricultural universities, beginning with the

Gobind Ballabh Pant University of Agriculture and Technology in Pantnagar in Uttar Pradesh, were gradually set up in all the states. The Land Grant model entailed linking research with extension, and taking research from the laboratories to the farms through an extension network linked with the university. In 1961, the government launched the Intensive Agriculture District Programme (IADP) to enhance productivity in the irrigated areas. Although the aim was to introduce improved seed along with a package of agronomic practices to take advantage of the available irrigation, the programme did not come up to expectations for want of high-yielding varieties of wheat.

The tall Indian wheat varieties did not provide an economic response to the application of chemical fertilisers and irrigation water, which agricultural scientists were desperately looking for. These had a tendency to fall down or lodge when fertilisers were applied, as a result of which the yields stagnated at less than one tonne per hectare. Even though breeding for improved non-lodging wheat varieties was initiated in the '50s, it did not meet with much success.

The missing link was provided by Dr. Norman Borlaug, who was then working with CIMMYT in Mexico. A few dwarf wheat strains of spring wheat that he sent to the Indian Agricultural Research Institute in New Delhi, looked impressive. In the next two to three years, a wide range of dwarf material was tested under the All India Wheat Research Project. In August 1964, the then Professor and Head of Department of Plant Breeding and Genetics at the IARI, Dr. M S Swaminathan, who later emerged as the architect of India's Green Revolution, proposed the launching of a National Demonstration Programme in farmers' fields. This would verify the results obtained in research plots and introduce farmers to new opportunities opened up by semi-dwarf varieties for considerably improving the productivity.

Says Swaminathan, "When small farmers, who with the help of scientists organised the National Demonstration Programme, harvested over 5 tonnes of wheat per hectare, its impact on the minds of other farmers was electric. The clamour for seeds began and the area under high-yielding varieties rose from 4 hectares in 1963-64 to over 4 million hectares in 1971-72. A small government programme became a mass movement." Subsequently, the government imported 18,000 tonnes of semi-dwarf wheat from Mexico in 1966 to be distributed in time for immediate sowing by farmers in the northwestern regions.

The seeds of the Green Revolution were, therefore, truly sown in the mid-'60s. What followed next is already part of contemporary history. Wheat production rose to 17 million tonnes in 1968, an increase of 5 million tonnes over the highest of 12 million tonnes harvested in 1964. Indira Gandhi officially recorded the impressive strides in agriculture by releasing a special stamp entitled 'Wheat Revolution' in July 1968. The success of wheat was later replicated in rice. Productivity increases were also recorded in cotton, sugarcane, millets and oilseeds. The famine-avoidance strategy that India launched had all the necessary ingredients for a sustained effort. In 1965, the Agricultural Prices Commission was constituted, with the basic objective of assuring fair prices for farm produce. The Commission, an autonomous body, works out the cost of cultivation for 22 agricultural commodities, and suggests a price to the government that includes profit that should serve as an incentive to grow more.

The administered price of wheat, for instance, was raised by 71 per cent between 1972 and 1981. Similarly, the price of rice was jacked up by 81 per cent during the same period. Later, to sustain the tempo of the rice revolution, the relative price structure was moved in its favour. In the early phase of the Green Revolution, the price policy favoured wheat, but after the mid-'70s it tilted in favour of rice. For oilseeds and pulses, the reverse was put into action. Relatively high support prices were announced in the early-'90s to make their cultivation profitable.

Perhaps the most significant aspect of the price support mechanism has been the insulation of farmers against a decline in prices. Wheat production was very high in 1971-72, as a consequence of which prices should have declined by at least 25 per cent. Since the government purchased all the wheat that was offered, farmers were actually paid 7 per cent more. Again, in 1982-83, when wheat production was 14.3 per cent higher than the previous year, the prices should have actually declined by 32.5 per cent. With the government purchasing all the wheat that flowed into the markets, however, the farmers realised 18 per cent more.

The biggest achievement of the Indian food policy, dependent more and more on buffer stocking and operational stock holding, has been the avoidance of famine-like conditions. It was with the basic objective of curbing consumption and ensuring an equitable distribution of available food supplies, especially in the deficit areas and among the poorer strata of society, that the Public Distribution System (PDS) was introduced. PDS supplies were also used for the 'food for work' programme as well as other anti-poverty programmes. At present, PDS covers more than 80 million families and the total foodgrains distributed through a network of 40,000 fair price shops accounts for over one-third of the total trade in foodgrain. The Green Revolution was certainly not a stroke of luck. The resilience was the result of progressive farm policies, determination, sustained efforts of scientists and the vigour of the Indian farmers who readily adopted the improved technology. This is what the Rockefeller Foundation had to say about the magnificent strides taken: "... The speed with which India's farmers and scientists suddenly gave their country an approach to an abundant food supply, has never been duplicated on an equal scale anywhere in the world, including the scientifically sophisticated United States."

Since the early-'70s, the country has avoided famine even during adverse weather conditions, through a carefully designed food security system involving the maintenance of both substantial grain reserves and an extensive PDS. The devastating drought of 1987, rated as the most severe of this century, affected nearly 155 million hectares. In spite of this, no starvation and hunger deaths were reported. This was essentially the result of excellent food management based on foodgrain reserves built largely from homegrown wheat and rice.

From the late-'60s, the rate of growth in food production generally exceeded the rate of growth of population. Despite the continuous growth in population, the per capita availability of foodgrain increased from 157 kg per year in 1955 to 177 kg per year in 1995, an increase of 15.7 per cent. Food production has risen roughly four times since the days of Independence, from 50 million tonnes in 1950-51 to 198 million tonnes in 1996-97. At the same time, the population has risen from 300 million to 960 million. Milk production too has risen to 68 million tonnes, the highest in the world. Food self-sufficiency was achieved in the late-'70s with the maintenance of a sizeable food buffer. Lately, India has begun to export foodgrain. About 5 million tonnes of rice and wheat were exported during the last few years. Skimmed milk powder amounting to about 5,000 tonnes has also been exported.

The increase in foodgrains was, however, disproportionate. The highest rate of growth was achieved in Punjab and Haryana, where, at a compound growth rate of 4.63 per cent, foodgrain production jumped from 7.23 million tonnes in 1964-65 to reach an all-time high of 30.33 million tonnes in 1995-96. Uttar Pradesh recorded a compound growth rate of 3.01 per cent. Production in Maharashtra, Madhya Pradesh, Bihar, Orissa and the northeastern states continues to stagger. Tamil Nadu and Andhra Pradesh, on the other hand, have recorded significant increases in rice yields, with production rising to 9.17 million tonnes and 11.58 million tonnes, respectively. The Green Revolution was aimed at the better-endowed regions. For millions of farmers languishing in the drylands, constituting more than 70 per cent of the cultivable lands, it continues to be a futile struggle. Despite emphasis on dryland farming during the past several decades, the scenario still remains grim. The undulating topography and the irregular rainfall pattern have combined to aggravate the situation. That the drylands produce about 42 per cent of the country's food shows that the future of farming lies in these areas. Nearly 83 per cent of sorghum, 81 per cent of pulses and 90 per cent of oilseeds grown in the country come from these areas. The poor yields and the fluctuations in production are indications of the scant attention drylands have received from policymakers and planners.

The problem of increasing productivity on drylands has serious socio-economic implications. With every passing year, the gap between the farmer's yields in irrigated areas and in the dry farming region is widening. One year of drought is enough to push a farmer into a deep well of poverty for another two to three years. Drought is a recurring phenomenon in arid and semi-arid areas. Fifty years after Independence, life for millions of people somehow surviving in the drylands continues to be worse than before.Thirty years after the dawn of the Green Revolution, Indian farmers are realising that their love affair with intensive agriculture is on the decline. Despite a bountiful monsoon for the tenth year in a row, harvests are not as plentiful as expected. However, with no let-up in the ever-increasing rise in population, Indian agriculture is once again at the crossroads.

Alarm bells have been ringing for quite some time. For nearly a decade, agricultural production had stagnated. The spectacular yield growth recorded in the post-Green Revolution years in Punjab and Haryana has receded into history. Of the multiple problems confronting agriculture, rapid fragmentation of land holdings is keeping pace with increasing population. In 1976-77, the average size of the holdings was estimated at 2 hectares, while in 1980-81, it came down to 1.8 hectares. Today, it stands at a mere 0.2 hectares. The total number of land holdings in 1981 were around 89 million; today these have crossed 100 million. By the turn of the century, the average land holding will come down to 0.11 hectares. It is quite obvious that with such small land holdings, Indian agriculture cannot adopt high-tech farm practices.

Basking in the afterglow of the Green Revolution, farming and agriculture have ceased to attract serious attention. In 1995-96, foodgrain production slumped to minus 3.60 per cent. In 1997-98, it further declined to minus 3.70 per cent, the worst-ever since the heady days of the Green Revolution. Foodgrain production in the frontline agricultural states of Punjab, Haryana and western Uttar Pradesh, comprising the country's food bowl, has decelerated. The miracle that began with wheat was replicated in rice. This is where the cost-benefit ratio has turned negative.

The downward spiral in food production continues through the southern regions of the country. Tamil Nadu, another Green Revolution area, is under tremendous strain from intensive cultivation. In Karnataka, the negative trend in yield levels of all food crops, barring cotton and sugarcane, are all too apparent. Farming in Karnataka can be clearly separated in two distinct classes, the 'creamy layer' of corporate agriculture occupying the fertile and irrigated areas, and the remaining low productive tracts at the mercy of the rains, constituting the tiny and small land holdings.

Kerala too, picturesque and verdant with tropical forests, is passing through its worst-ever crisis on the foodgrain front. As if this is not enough, the number of landless in rural areas too has been multiplying over the past few decades, at an estimated rate of two million every year. The rapid expansion in the number of land holdings, with less than a hectare under plough, has meant an additional 6,00,000 farmers every five years or so. The negative terms of trade for agriculture and the declining public sector investments in farming are indices of the sluggishness in this sector.

The Economic Survey for 1996-97 has further sounded a warning. The compound growth rate in foodgrain production at 1.7 per cent between 1991 and 1996, is lower than the annual population growth of 1.9 per cent. The survey adds that even a marginal fall of three to four per cent in foodgrain output can cause the prices of essential commodities to escalate sharply, necessitating government intervention by way of resorting to imports. This happened in 1993 and again in 1996.

The philosophy of agricultural planning is changing. Gone are the days when the nation's emphasis was solely on attaining self-sufficiency in foodgrain production. Gone also are the days when farmers were newly independent India's heroes, revered for their role in keeping hunger and starvation at bay. Today, at a time when food production struggles to keep pace with the burgeoning population growth, farmers are being asked to diversify, produce crops that are suitable for export and to compete in the international market. With the promise of cheap food available off the shelf in the global market, the focus has shifted from agriculture to industry, trade and commerce, from the small and marginal farmers to the agri-processing companies, which alone can bring in investments and add value to produce.

The restructuring of the Indian economy too has put the emphasis on agri-exports. Cultivation of staple food is being replaced by cash crops, tomatoes in place of wheat, durum wheat (for bakery purposes) replacing wheat as a staple diet in Punjab and Haryana, flowers in place of rice, and so on. In the coastal areas, private enterprise is taking away the fish catch, depriving the local communities of a livelihood and their only source of nutrition. In Kerala, vast tracts of forests and paddy fields have been converted into rubber, coffee and coconut plantations. Every year, about 25,000 hectares of good paddy land are being diverted for non-paddy purposes. The structural transformation is not only peculiar to Kerala. It is happening in almost all the states. Commercial crops are eating into the fertile land tracts meant for growing essential foodgrains, thereby making the states rely more heavily on the PDS. The diversion of good agricultural land, which in any case is limited, to commercial farming and even industries, is further restricting the ability to grow enough foodgrain.

SUSTAINABLE AGRICULTURE: CLIMATE CHANGE AND FOOD SECURITY

The exponential population increase in recent decades has increased the practice of agricultural land conversion to meet demand for food which in turn has increased the effects on the environment. The global population is still increasing and will eventually stabilise, as some critics doubt that food production, due to lower yields from global warming, can support the global population. Organic farming is a multifaceted sustainable agriculture set of practices that can have a lower impact on the environment Other specific methods include: permaculture; and biodynamic agriculture which incorporating a spiritual element.

The links between climate change and food security have, to date, largely been explored in relation to impacts on crop productivity and hence, food production. For instance, Gregory *et al.* (1999) summarized experimental findings on wheat and rice that indicated decreased crop duration (and hence yield) of wheat as a consequence of warming and reductions in yields of rice of about 5% $°C^{-1}$ rise above 32 °C. These effects of temperature were considered sufficiently detrimental that they would largely offset any increase in yield as a consequence of increased atmospheric carbon dioxide (CO_2) concentration. Several reviews have further assessed the potential consequences of changes in climate on the growth and yield of crop plants, concluding that the earlier-anticipated benefits of CO_2 fertilization would be largely offset by nutrient limitations, pollutants and further interactions with climatic factors.

Similarly, simulations of maize production in Africa and Latin America using climate data from the HadCM2 model to generate characteristic daily weather data for 2055 predicted an overall reduction of 10%. This, and other similar projections, uses a process-based crop production model (such as the crop environment resource synthesis family) to link climate to plant physiological processes. Yield can then be modelled for a uniform crop and upscaled to a larger area normally within some form of geographic information system (GIS). The aggregate result of Jones & Thornton (2003), though, hides considerable variability within and between countries, and, as they point out, also ignores the fact that maize is commonly used as fodder as well as food as part of a complex production system.

The spatial variation in effects of climate change was an explicit component of the study on potential yields of rainfed cereal crops undertaken by Fischer *et al.* (2001). In this analysis, climate predictions in 2080 were obtained from various global circulation models (GCMs) and assessments made based on current populations and socio-economic conditions.

The output demonstrated that cereal producing regions of Canada, and northern Europe and Russia might be expected to increase production as a consequence of the climate changes predicted by GCMs, while many parts of the world would suffer losses including the western edge of the USA prairies, eastern Brazil and western Australia. Overall, the results of this and subsequent work that included assessments of future populations and alternative future socio-economic conditions, demonstrated that climate change would benefit the developed countries more than the developing countries even if cropping practices evolved to allow more than one rainfed crop per year. Moreover, the anticipated demographic growth and socio-economic development in these developing countries would result in substantial increases in food requirements thereby exacerbating the detrimental effects of climate change.

The spectre of climate change, together with other global environmental changes such as changes in water availability, and land cover, and altered nitrogen availability and cycling (all strongly influenced by human activities), has increased concerns about achieving food security especially for poor people. There is also concern that meeting the global demand for food resulting from higher population and changing dietary preferences will further degrade the environment both through additional destruction of native vegetation and increased intensification of cropped areas. This may, in turn, further undermine the food systems upon which food security is based.

While there has been considerable progress in understanding the sensitivities of crop yield to climate change, assessments of climate change effects on food security remain rather limited. Food security is concerned not only with food availability but also with access to and utilization of, food so that studies, which focus only on crop production provide only a partial assessment of food security–climate change relationships. This is of particular concern as many of the policy-based issues that are relevant to the international and sustainable development communities are posed in the context of food security prospects. Engagement of these policy communities requires a much broader and comprehensive research framework.

Recognition that food security depends on robust food systems that encompass issues of availability, access and utilization (not merely production alone), and consequently that the nature of key research issues changes as questions more related to food security are formulated, led the international global change research community to establish the joint project global environmental change and food systems (GECAFS) in 2001. GECAFS is an interdisciplinary research programme that aims to improve understanding of the relationship between food systems and the earth system to deliver science-based tools for analysing socio-economic and environmental consequences of strategies seeking to adapt to climate and other environmental changes. The purposes of this chapter are to: (i) outline the notion of food systems, their relation to food security, and links to climate change; (ii) explore the vulnerability of food systems to environmental change; (iii) examine some prospects for adaptation of food systems in response to climate change and (iv) highlight some of the environmental feedbacks resulting from modifications to food systems.

FOOD SECURITY, FOOD SYSTEMS AND THE LINK TO CLIMATE

There are several definitions of what constitutes food systems each formulated in relation to a specific range of issues. For GECAFSs purpose, food systems are defined as a set of dynamic interactions between and within the biogeophysical and human environments which result in the production, processing, distribution, preparation and consumption of food. They encompass components of: (i) food availability (with elements related to production, distribution and exchange); (ii) food access (with elements related to affordability, allocation and preference) and (iii) food utilization. Food systems, then, involve much broader considerations than productivity and production alone. They underpin food security, which is the state achieved when food systems operate such that 'all people, at all times, have physical and economic access to sufficient, safe and nutritious food to meet their dietary needs and food preferences for an active and healthy life' (FAO 1996). Food security is diminished when food systems are stressed. This can be caused by a range of factors in addition to climate and other environmental changes (*e.g.* conflict, changes in international trade agreements and policies, HIV/AIDS) and may be particularly severe when these factors act in combination. Access to culturally acceptable food by individuals and communities, or means for its procurement, is increasingly being elaborated into human rights legislation.

Food systems may be simple, as in the case of a subsistence farmer who produces, processes and consumes food on farm. However, there are comparatively few individuals or households in the world that are totally self-reliant for food throughout a year, and in almost all cases there is an element of bartering, exchange, or the cash economy to bring food into the household. In many places, the food system has changed radically in the last century and continues to become increasingly complex. The intensification of agricultural production, since the 1940s has been accompanied by profound changes in the organization of food systems around the world including changes in distribution, marketing, affordability and preferences for particular food items. These changes are especially obvious in the USA and Europe, where market globalization has occurred with global sourcing of products by retailers direct from producers in the case of fresh fruit and vegetables and from a few, large manufacturers for other food products.

Such changes have also shifted economic and political power from farmers to retailers, from national legislative bodies to regional and global organizations, and from the state to multinational corporations. For example, in a study of selected OECD countries in Europe, Grievink (2003) determined that the food chain has some 160 million consumers of whom about 3.2 million are also farmers or food producers, but the link between these groups is increasingly determined by the small number of food processors/manufacturers (about 90 000) and the even smaller number of buyers (about 100) for the supermarket chains.

Components of the food chain in selected OECD countries of Europe indicating the two inverted pyramid structure that relates farmers to consumers.

(From Grievink 2003.) Food systems around the world are changing very rapidly as urbanization and globalization proceed apace. The urbanization of many predominantly rural countries in the last three decades has been accompanied by the rapid growth of supermarkets in many, often accompanied by foreign investment by global retail chains.

However, even in poor countries such as Kenya, where per capita gross domestic product (GDP) was less than \$400 yr^{-1} in 2002, supermarkets have grown from a tiny niche market in 1997 to be greater than 20% of urban food retailing today. This growth has been a consequence of three major factors. First, rapid urbanization has seen the proportion of the population based in urban areas grow from 0.13 in 1975 to 0.36 in 2000 with an expectation of greater than 0.5 in 2013; second, trade and domestic market liberalization, since 1993 has seen the removal of import controls and the deregulation of prices and third, movement towards price-based competition between the indigenous chains. Success within Kenya is now spreading to other East African countries with important effects on the market conditions faced by farmers including the decline of traditional wholesalers (and the smallholder producers from whom they buy) and the increase in direct purchases from larger farms.

These marked changes in access and utilization of food around the world provide a context in which to evaluate the likely effects of climate and other environmental changes on crop production and food security. Not all food systems or parts of food systems are equally vulnerable to environmental changes because the capacity to cope with existing variability in bio-physical and socio-economic systems, and the ability of humans to perceive environmental changes and to adapt food systems, differs. Human vulnerability includes both the likelihood of exposure to stresses as well as the capacity to cope with such stresses (*i.e.* sensitivity). Vulnerability and poverty are often inter-related because both the likelihood of exposure to stresses is greater among the poor and because a large proportion of their resources are spent either purchasing or producing food, thereby reducing their capacity to cope with perturbations. Moreover, pursuit of food security frequently involves trade-offs with expenditure on health and education reducing still further the ability to improve longer-term living conditions or resilience to stress and shock. Food insecurity is experienced at a range of spatial scales from individual households to regions, as well as a range of time-scales, and reductions of vulnerability at one scale do not automatically flow to the next scale. For example, regional policy decisions do not always convert to successful local implementation especially if distribution services are inadequate, or food preferences are ignored.

Climate variation is one of several interacting factors that affect food security. For example, in studies of household food security in southern Africa, climate/environment was only one of some 33 drivers mentioned as important by householders. The mix of drivers varied across the region but in all communities many interacting factors resulted in vulnerability to food shortages. Overall, however, climate/environment was one of the seven factors influencing food security that were frequently cited, because of its role both as an ongoing issue (57% of cases, where it was mentioned) and as a 'shock' (43%). The impacts of sudden shocks such as drought are felt, then, on top of ongoing long-term stresses, and the low ability to cope with such shocks and to mitigate long-term stresses means that the employment of coping strategies that might be available to others, is at a too high cost or, simply, unavailable. Typically, reliance on purchased food increases in drought years due to losses in food production leading to an increase in poverty due to the synergistic action of other drivers such as rising food prices and unemployment. Scholes & Biggs (2004) record that the food security crisis in southern Africa in 2002–2003 was not simply a result of drought alone and, indeed, climatic stress was not as severe as in previous crises. Maize production during the preceding growing season was only 5.5% less than the previous five year average so food stocks at the start of the climatic shock were not unusually low. It was concluded that the crisis was indicative of entrenched vulnerability resulting from a range of regional and global political and economic factors including high food prices, legacies of structural adjustment, government policies, conflict and war, policies on genetically modified foods, and poor responses to the HIV/AIDS pandemic. The key here was that conditions, which weakened food systems in the region were already in place. The moderate climatic shock intensified food insecurity and the long-term vulnerability of the region.

The seven most frequently cited drivers in 49 studies of household-level food insecurity in southern Africa. The numbers in the arrows indicate the number of citations, as a percentage of 555 citations of 33 possible drivers. The drivers shaded in grey were noted as being chronic, while those in white indicate drivers experienced mainly as 'shocks'. The shaded arrows indicate drivers that acted primarily via reductions in food production, while the white arrows indicate those which acted by restricting access to food.

QUANTIFYING THE IMPACTS ON FOOD SECURITY

A number of studies have recently quantified the impacts of climate change on food security. In terms of quantifying agronomic yield change projections, these studies are either based on the AEZ tools developed by the IIASA analysis or the Decision Support System for Agrotechnology Transfer suite of crop models; all use the IIASA-Basic Linked System (BLS) economic model for assessing economic impacts. These tools, with some modifications relating to how crop yield changes are simulated, have also been used by others to undertake similar assessments and provide sensitivity analyses across a range of SRES and general circulation model (GCM) projections.

Many other simulations have also examined the effects of climate change with and without adaptation (induced technological progress, domestic policy change, international trade liberalization, etc.), with and without mitigation (*e.g.*, CO_2 stabilization, variants for temperature, rainfall change and distribution) or provide impact assessments for different speeds of climate change.

The quantitative results for food security, trying to illuminate some of the differences and extract the main messages that emerge from the various studies. Unless indicated, all simulation results discussed below include the combined effects of climate change and elevated CO_2 on crops. The key messages can be summarized as follows:

First, it is very likely that climate change is likely to increase the number of people at risk of hunger compared with reference scenarios with no climate change; the exact impacts will, however, strongly depend on the projected socio-economic developments. For instance, it is estimated that climate change would increase the number of undernourished in 2080 by 5–26%, compared with no climate change or by between 5 million and 10 million (B1 SRES) and 120 million to 170 million people (A2 SRES), with within-SRES ranges depending on GCM climate projections. Using only one GCM scenario, others projected small reductions by 2080, *i.e.*, –5%, or –10 million (B1) to –30 million (A2) people, and slight increases of +13%–26%, or H–10 million (B2) to 30 million (A1) people.

Second, it is likely that the magnitude of these climate impacts will be small compared with the impact of socio-economic development. The limitations of socio-economic forecasts, these studies suggest that robust economic growth and a decline in population growth projected for the 21st century will, in all but one scenario (SRES A2), significantly reduce the number of people at risk of hunger in 2080. At any rate, the prevalence of undernourishment will decline as all scenarios assume that world population will continue to grow to 2080, albeit at lower rates. Compared with FAO estimates of 820 million undernourished in developing countries today, several studies estimate reductions of >75% by 2080, or H–560 million to 700 million people, projecting 100 million to 240 million undernourished by 2080 (A1, B1, and B2).

As mentioned, the only exception is scenario A2, where the number of the hungry is forecast to decrease only slightly by 2080; but the higher population growth rates in A2 compared with other scenarios mean that the prevalence of undernourishment will decline drastically. However, these analyses also confirm that the progress will be unevenly distributed over the developing world, and more importantly progress will be slow during the first decades of the outlook. With or without climate change, the millennium development goal of halving the prevalence of hunger by 2015 is unlikely to be realized before 2020–2030.

In addition to socio-economic pressures considered within the IPCC SRES scenarios, food production may increasingly compete with bio-energy in coming decades; studies addressing possible consequences for world food supply have only started to surface, providing both positive and negative views. Importantly, none of the major world food models discussed herein have yet considered such competition.

Third, sub-Saharan Africa is likely to surpass Asia as the most food-insecure region. However, this is largely independent of climate change and is mostly the result of the socio-economic development paths assumed for the different developing regions in the SRES scenarios. Throughout most SRES and climate-change scenarios sub-Saharan Africa accounts for 40–50% of global hunger by 2080, compared with H–24% today; in some simulations sub-Saharan Africa even accounts for 70–75% of global undernourishment by 2080. These high estimates have emerged from slower growth variants of the A2 and B2 scenarios; also an A2 variant with slower population growth yields a sharper concentration of hunger in sub-Saharan Africa. For regions other than sub-Saharan Africa, results largely depend on GCM scenarios and therefore are highly uncertain.

Fourth, although there is significant uncertainty on the effects of elevated CO_2 on crop yields, this uncertainty is carried to a much lesser extent on food security. This result emerges from a comparison of climate change simulations with and without CO_2 fertilization effects on crop yields.

As can be seen, higher CO_2 fertilization would not greatly affect global projections of hunger. In view of the fact that essentially all SRES worlds are characterized by much higher real incomes, much improved transportation and communication options, and still sufficient global food production, the somewhat smaller supplies will not be able to make a dent in global food security outcomes.

Many studies find that climate change without CO_2 fertilization would reduce the number of undernourished people by 2080 only by some 20 million to 140 million (120 million to 380 million for SRES A1, B1, and B2 without, compared with 100 million to 240 million with CO_2 fertilization effect). The exception again in these studies is SRES A2, under which the assumption of no CO_2 fertilization results in a projected range of 950 million to 1,300 million people undernourished in 2080, compared with 740 million to 850 million projected under climate change, but with CO_2 effects on crops.

Finally, recent research suggests large positive effects of climate stabilization for the agricultural sector. However, as the stabilizing effects of mitigation measures can take several decades to be realized from the moment of implementation, the benefits for crop production may be realized only in the second half of this century. Importantly, even in the presence of robust global long-term benefits, regional and temporal patterns of winners and losers that can be projected with current tools are highly uncertain and depend critically on the underlying GCM projections. The finding that socio-economic development paths have an important bearing on future food security and that they are likely to top the effects of climate change should not, or at least not only, be interpreted as a probability-based forecast. This is because SRES scenarios offer a range of possible outcomes "without any sense of likelihood". Yet SRES scenarios, like all scenarios, do not overcome the inability to accurately project future changes in economic activity, emissions, and climate.

Second, the existing global assessments of climate change and food security have only been able to focus on the impacts on food availability and access to food, without quantification of the likely important climate change effects on food safety and vulnerability (stability). This means that these assessments neither include potential problems arising from additional impacts due to extreme events such as drought and floods nor do they quantify the potential impacts of changes in the prevalence of food-borne diseases (positive as well as negative) or the interaction of nutrition and health effects through changes in the proliferation of vector-borne diseases such as malaria. On the food availability side, they also exclude the impacts of a possible sea-level rise for agricultural production or those that are associated with possible reductions of marine or freshwater fish production.

Third, it is important to note that even in terms of food availability, all current assessments of world food supply have focused only on the impacts of mean climate change, *i.e.*, they have not considered the possibility of significant shifts in the frequency of extreme events on regional production potential, nor have they considered scenarios of abrupt climate or socio-economic change; any of these scenario variants is likely to increase the already negative projected impacts of climate change on world food supply. Models that take into account the specific biophysical, technological, and market responses necessary to simulate realistic adaptation to such events are not yet available.

Fourth, this review finds that recent global assessments of climate change and food security rest essentially on a single modeling framework, the IIASA system, which combines the FAO/IIASA AEZ model with various GCM models and the IIASA BLS system, or on close variants of the IIASA system. This has important implications for uncertainty, given that the robustness of all these assessments strongly depends on the performance of the underlying models. There is, therefore, a clear need for continued and enhanced validation efforts of both the agro-climatological and food trade tools developed at IIASA and widely used in the literature.

Finally, we note that assessments that do not only provide scenarios, but also attach probabilities for particular outcomes to come true could provide an important element for improved or at least better-informed policy decisions.

A number of possibilities are offered to address the related modeling challenges. One option would be to produce such estimates with probability-based estimates of the (key) model parameters. Alternatively, the various scenarios could be constructed so that they reflect expert judgements on a particular issue. It would be desirable to attach probabilities to existing scenarios.

Information on how likely the suggested outcomes are would contribute greatly to their usefulness for policy makers and help justify (or otherwise) policy measures to adapt to or mitigate the impacts of climate change on food security. Climate change will affect all four dimensions of food security, namely food availability (*i.e.*, production and trade), access to food, stability of food supplies, and food utilization. The importance of the various dimensions and the overall impact of climate change on food security will differ across regions and over time and, most importantly, will depend on the overall socio-economic status that a country has accomplished as the effects of climate change set in.

Essentially all quantitative assessments show that climate change will adversely affect food security. Climate change will increase the dependency of developing countries on imports and accentuate existing focus of food insecurity on sub-Saharan Africa and to a lesser extent on South Asia. Within the developing world, the adverse impacts of climate change will fall disproportionately on the poor. Many quantitative assessments also show that the socio-economic environment in which climate change is likely to evolve is more important than the impacts that can be expected from the biophysical changes of climate change.

Less is known about the role of climate change for food stability and utilization, at least in quantitative terms. However, it is likely that differences in socio-economic development paths will also be the crucial determinant for food utilization in the long run and that they will be decisive for the ability to cope with problems of food instability, be they climate-related or caused by other factors.

Finally, all quantitative assessments we reviewed show that the first decades of the 21st century are expected to see low impacts of climate change, but also lower overall incomes and still a higher dependence on agriculture. During these first decades, the biophysical changes as such will be less pronounced but climate change will affect those particularly adversely that are still more dependent on agriculture and have lower overall incomes to cope with the impacts of climate change. By contrast, the second half of the century is expected to bring more severe biophysical impacts but also a greater ability to cope with them. The underlying assumption is that the general transition in the income formation away from agriculture towards nonagriculture will be successful.

How strong the impacts of climate change will be felt over all decades will crucially depend on the future policy environment for the poor. Freer trade can help to improve access to international supplies; investments in transportation and communication infrastructure will help provide secure and timely local deliveries; irrigation, a promotion of sustainable agricultural practices, and continued technological progress can play a crucial role in providing steady local and international supplies under climate change.

CLIMATE CHANGE AFFECTS FOOD SECURITY AND AGRICULURE

Climate change affects agriculture and food production in complex ways. It affects food production directly through changes in agro-ecological conditions and indirectly by affecting growth and distribution of incomes, and thus demand for agricultural produce. Impacts have been quantified in numerous studies and under various sets of assumptions. A selection of these results is presented in *Quantifying the Impacts on Food Security*. Here it is useful to summarize the main alterations in the agro-ecological environment that are associated with climate change.

Changes in temperature and precipitation associated with continued emissions of greenhouse gases will bring changes in land suitability and crop yields. In particular, the Intergovernmental Panel on Climate Change (IPCC) considers four families of socio-economic development and associated emission scenarios, known as Special Report on Emissions Scenarios (SRES) A2, B2, A1, and B1. Of relevance to this review, of the SRES scenarios, A1, the "business-as-usual scenario," corresponds to the highest emissions, and B1 corresponds to the lowest. The other scenarios are intermediate between these two. Importantly for agriculture and world food supply, SRES A2 assumes the highest projected population growth of the four (United Nations high projection) and is thus associated to the highest food demand. Depending on the SRES emission scenario and climate models considered, global mean surface temperature is projected to rise in a range from 1.8°C (with a range from 1.1°C to 2.9°C for SRES B1) to 4.0°C (with a range from 2.4°C to 6.4°C for A1) by 2100. In temperate latitudes, higher temperatures are expected to bring predominantly benefits to agriculture: the areas potentially suitable for cropping will expand, the length of the growing period will increase, and crop yields may rise.

A moderate incremental warming in some humid and temperate grasslands may increase pasture productivity and reduce the need for housing and for compound feed. These gains have to be set against an increased frequency of extreme events, for instance, heat waves and droughts in the Mediterranean region or increased heavy precipitation events and flooding in temperate regions, including the possibility of increased coastal storms; they also have to be set against the fact that semiarid and arid pastures are likely to see reduced livestock productivity and increased livestock mortality.

In drier areas, climate models predict increased evapotranspiration and lower soil moisture levels. As a result, some cultivated areas may become unsuitable for cropping and some tropical grassland may become increasingly arid. Temperature rise will also expand the range of many agricultural pests and increase the ability of pest populations to survive the winter and attack spring crops.

Another important change for agriculture is the increase in atmospheric carbon dioxide (CO_2) concentrations. Depending on the SRES emission scenario, the atmospheric CO_2 concentration is projected to increase from H"379 ppm today to >550 ppm by 2100 in SRES B1 to >800 ppm in SRES A1FI. Higher CO_2 concentrations will have a positive effect on many crops, enhancing biomass accumulation and final yield. However, the magnitude of this effect is less clear, with important differences depending on management type (*e.g.*, irrigation and fertilization regimes) and crop type. Experimental yield response to elevated CO_2 show that under optimal growth conditions, crop yields increase at 550 ppm CO_2 in the range of 10% to 20% for C_3 crops (such as wheat, rice, and soybean), and only 0–10% for C_4 crops such as maize and sorghum. Yet the nutritional quality of agricultural produce may not increase in line with higher yields. Some cereal and forage crops, for example, show lower protein concentrations under elevated CO_2 conditions.

Finally, a number of recent studies have estimated the likely changes in land suitability, potential yields, and agricultural production on the current suite of crops and cultivars available today. Therefore, these estimates implicitly include adaptation using available management techniques and crops, but excluding new cultivars from breeding or biotechnology. These studies are in essence based on the FAO/International Institute for Applied Systems Analysis (IIASA) agro-ecological zone (AEZ) methodology. For instance, pioneering work in ref. 9 suggested that total land and total prime land would remain virtually unchanged at the current levels of 2,600 million and 2,000 million hectares (ha), respectively.

The same study also showed, however, more pronounced regional shifts, with a considerable increase in suitable cropland at higher latitudes (developed countries +160 million ha) and a corresponding decline of potential cropland at lower latitudes (developing countries "110 million ha). An even more pronounced shift within the quality of cropland is predicted in developing countries. The net decline of 110 million ha is the result of a massive decline in agricultural prime land of H"135 million ha, which is offset by an increase in moderately suitable land of >20 million ha. This quality shift is also reflected in the shift in land suitable for multiple cropping. In sub-Saharan Africa alone, land for double cropping would decline by between 10 million and 20 million ha, and land suitable for triple copping would decline by 5 million to 10 million ha. At a regional level, similar approaches indicate that under climate change, the biggest losses in suitable cropland are likely to be in Africa, whereas the largest expansion of suitable cropland is in the Russian Federation and Central Asia. Global and regional weather conditions are also expected to become more variable than at present, with increases in the frequency and severity of extreme events such as cyclones, floods, hailstorms, and droughts. By bringing greater fluctuations in crop yields and local food supplies and higher risks of landslides and erosion damage, they can adversely affect the stability of food supplies and thus food security.

Neither climate change nor short-term climate variability and associated adaptation are new phenomena in agriculture, of course. As shown, for instance, in ref. 9, some important agricultural areas of the world like the Midwest of the United States, the northeast of Argentina, southern Africa, or southeast Australia have traditionally experienced higher climate variability than other regions such as central Africa or Europe. They also show that the extent of short-term fluctuations has changed over longer periods of time. In developed countries, for instance, short-term climate variability increased from 1931 to 1960 as compared with 1901 to 1930, but decreased strongly in the period from 1961 to 1990. What is new, however, is the fact that the areas subject to high climate variability are likely to expand, whereas the extent of short-term climate variability is likely to increase across all regions. Furthermore, the rates and levels of projected warming may exceed in some regions the historical experience.

If climate fluctuations become more pronounced and more widespread, droughts and floods, the dominant causes of short-term fluctuations in food production in semiarid and subhumid areas, will become more severe and more frequent. In semiarid areas, droughts can dramatically reduce crop yields and livestock numbers and productivity. Again, most of this land is in sub-Saharan Africa and parts of South Asia, meaning that the poorest regions with the highest level of chronic undernourishment will also be exposed to the highest degree of instability in food production.

How strongly these impacts will be felt will crucially depend on whether such fluctuations can be countered by investments in irrigation, better storage facilities, or higher food imports. In addition, a policy environment that fosters freer trade and promotes investments in transportation, communications, and irrigation infrastructure can help address these challenges early on. Climate change will also affect the ability of individuals to use food effectively by altering the conditions for food safety and changing the disease pressure from vector, water, and food-borne diseases. The IPPC Working Group II provides a detailed account of the health impacts of climate change of its fourth assessment report. It examines how the various forms of diseases, including vector-borne diseases such as malaria, are likely to spread or recede with climate change. This article focuses on a narrow selection of diseases that affect food safety directly, *i.e.*, food and water-borne diseases.

The main concern about climate change and food security is that changing climatic conditions can initiate a vicious circle where infectious disease causes or compounds hunger, which, in turn, makes the affected populations more susceptible to infectious disease. The result can be a substantial decline in labour productivity and an increase in poverty and even mortality. Essentially all manifestations of climate change, be they drought, higher temperatures, or heavy rainfalls have an impact on the disease pressure, and there is growing evidence that these changes affect food safety and food security.

The recent IPCC report also emphasizes that increases in daily temperatures will increase the frequency of food poisoning, particularly in temperate regions. Warmer seas may contribute to increased cases of human shellfish and reef-fish poisoning (ciguatera) in tropical regions and a poleward expansion of the disease. However, there is little new evidence that climate change significantly alters the prevalence of these diseases.

Several studies have confirmed and quantified the effects of temperature on common forms of food poisoning, such as salmonellosis. These studies show an approximately linear increase in reported cases for each degree increase in weekly temperature. Moreover, there is evidence that temperature variability affects the incidence of diarrhoeal disease. A number of studies found that rising temperatures were strongly associated with increased episodes of diarrhoeal disease in adults and children. These findings have been corroborated by analyses based on monthly temperature observations. Several studies report a strong correlation between monthly temperature and diarrhoeal episodes on the Pacific Islands, Australia, and Israel.

Extreme rainfall events can increase the risk of outbreaks of water-borne diseases particularly where traditional water management systems are insufficient to handle the new extremes. Likewise, the impacts of flooding will be felt most strongly in environmentally degraded areas, and where basic public infrastructure, including sanitation and hygiene, is lacking. This will raise the number of people exposed to water-borne diseases (*e.g.*, cholera) and thus lower their capacity to effectively use food.

Access to food refers to the ability of individuals, communities, and countries to purchase sufficient quantities and qualities of food. Over the last 30 years, falling real prices for food and rising real incomes have led to substantial improvements in access to food in many developing countries. Increased purchasing power has allowed a growing number of people to purchase not only more food but also more nutritious food with more protein, micronutrients, and vitamins. East Asia and to a lesser extent the Near-East/ North African region have particularly benefited from a combination of lower real food prices and robust income growth. From 1970 to 2001, the prevalence of hunger in these regions, as measured by FAO's indicator of undernourishment, has declined from 24% to 10.1% and 44% to 10.2% respectively. In East Asia, it was endogenous income growth that provided the basis for the boost in demand for food, which was largely produced in the region; in the Near-East North African region demand was spurred by exogenous revenues from oil and gas exports, and additional food supply came largely from imports. But in both regions, improvements in access to food have been crucial in reducing hunger and malnutrition.

FAO's longer-term outlook to 2050 suggests that the importance of improved demand side conditions will even become more important over the next 50 years. The regions that will see the strongest reductions in the prevalence of undernourishment are those that are expected to see the highest rates of income growth. South Asia stands to benefit the most. Spurred by high income growth the region is expected to reduce the prevalence of undernourishment from >22% currently to 12% by 2015 and just 4% by 2050. Progress is also expected for sub-Saharan Africa, but improvements will be less pronounced and are expected to set in later. Over the next 15 years, for instance, the prevalence of undernourishment will decline less than in other regions, from H"33% to a still worrisome 21%, as significant constraints (soil nutrients, water, infrastructure, etc.) will limit the ability to further increase food production locally, while continuing low levels of income rule out the option of importing food. In the long run, however, sub-Saharan Africa is expected to see a more substantial decline in hunger; by 2050, <6% of its total population are expected to suffer from chronic hunger. It is important to note that these FAO projections do not take into account the effects of climate change.

By coupling agro-ecologic and economic models, others have gauged the impact of climate change on agricultural gross domestic product (GDP) and prices. At the global level, the impacts of climate change are likely to be very small; under a range of SRES and associated climate-change scenarios, the estimates range from a decline of "1.5% to an increase of +2.6% by 2080. At the regional level, the importance of agriculture as a source of income can be much more important. In these regions, the economic output from agriculture itself will be an important contributor to food security. The strongest impact of climate change on the economic output of agriculture is expected for sub-Saharan Africa, which means that the poorest and already most food-insecure region is also expected to suffer the largest contraction of agricultural incomes. For the region, the losses in agricultural GDP, compared with no climate change, range from 2% to 8% for the Hadley Centre Coupled Model, version 3 and 7–9% for the Commonwealth Scientific and Industrial Research Organisation projections. Essentially all SRES development paths describe a world of robust economic growth and rapidly shrinking importance of agriculture in the long run and thus a continuation of a trend that has been underway for decades in many developing regions. SRES scenarios describe a world where income growth will allow the largest part of the world's population to address possible local production shortfalls through imports and, at the same time, find ways to cope with safety and stability issues of food supplies. It is also a world where real incomes rise more rapidly than real food prices, which suggests that the share of income spent on food should decline and that even high food prices are unlikely to create a major dent in the food expenditures of the poor. However, not all parts of the world perform equally well in the various development paths and not all development paths are equally benign for growth. Where income levels are low and shares of food expenditures are high, higher prices for food may still create or exacerbate a possible food security problem.

There are a number of studies that have ventured to measure the likely impacts of climate change on food prices. The basic messages that emerge from these studies are: first, on average, prices for food are expected to rise moderately in line with moderate increases of temperature (until 2050); some studies even foresee a mild decline in real prices until 2050. Second, after 2050 and with further increases in temperatures, prices are expected to increase more substantially.

In some studies and for some commodities (rice and sugar) prices are forecast to increase by as much as 80% above their reference levels without climate change. Third, price changes expected from the effects of global warming are, on average, much smaller than price changes from socio-economic development paths. For instance, the SRES A2 scenario would imply a price increase in real cereal prices by H"170%. The (additional) price increase caused by climate change (in the Hadley Centre Coupled Model, version 3, climate change case) would only be 14.4%. Overall, this appears to be the sharpest price increase reported and it is not surprising that this scenario would imply a stubbornly high number of undernourished people until 2080. However, it is also needless to say that a constant absolute number of undernourished people would still imply a sharp decline in the prevalence of hunger, and, given the high population assumptions in the SRES A2 world (>13.6 billion people globally and >11.6 billion in the developing world) this would imply a particularly sharp drop in the prevalence from currently 17% to H"7% by 2080.

IMPACT OF CLIMATE CHANGE IN INDIAN AGRICULTURE

The Food and Agricultural Organization (FAO) has warned that India could lose upto 125 million tones of cereals. The Intergovernmental Panel on Climate Change (IPCC) which was awarded Nobel Prize in the year 2008 has warned that due to global warming, the sea levels in Asia will rise by atleast 40cm by

2100 flooding vast areas on the coastline, including some of the most densely populated cities, whose population will be forced to migrate inland (The Hindu Business Line dated 03-11-2007). The first to be affected by the climate change will be those living in the fringe zones, working outdoors or whose livelihoods depend directly on the physical environment. Viewed from these perspectives, the main sectors facing risks from direct impact could be agriculture, water resources and health.

The IPCC report indicates high probability of crop losses with increase in temperature in the tropical regions. Studies do confirm this trend. Among cereal crops important for food security, wheat is most sensitive to even small increase in temperature. Relatively, rice has greater tolerance to temperature increases. Increasing climate variability could result in considerable season/annual fluctuation in food production. All agricultural commodities are subject to such variability. The most important contributor to global warming is the increase in atmospheric CO_2 levels due to human activity. The following are the some of the suggestions to prevent global warming.

CO_2 emission can be cut by reducing the use of fossil fuels by cutting back on car use, investing in energy efficiency, implementing energy conservation measures and utilizing renewable resources such as wind, solar and hydropower. Reducing fossil fuel user will, coincidentally, also reduce emissions of methane, nitrous oxide and ozone.

Since CO_2 is consumed by plants and trees (known as carbon-sinks), reversing deforestation and implementing reforestation programmes may reduce levels in the atmosphere. The need to take such measures was first recognized at the RIO climate change convention in 1992, after which over 160 countries pledged to limit emissions of CO_2 and to protect and enhance natural 'sinks' of CO_2. Trees play a unique role in the global carbon cycle. They are the largest land-based mechanism for removing CO_2 from the air. Trees are able to store a large amount of CO_2 in their structure. An acre of forest will absorb about ten times the CO_2 amount absorbed by an acre of crop land. Planting more trees and reducing more trees and reducing timber cut worldwide will help to restore the balance.

Try to cut down demand for electricity when we reduce electricity power use, we save money, breathe cleaner air and help to reduce the global warming problem. Government should also encourage the use of mass transit; provide tax rebates for people who buy low and no-pollution vehicles and subsidies to fossil fuel and the nuclear industries.

Green house gas emissions be managed using an incentive based policy, such as market based approach to capping and reducing such emissions. This type of strategy provides clear incentives for changes in business practices and the development of new technologies. It assures that economic forces are directed to keeping the cost of reducing emissions as low as they can be. Many industrial nations have now adopted policies intended to limit green house gases.

Try to follow the environment-friendly policies like 'Reduce', 'Reuse' and 'Recycle'. Save containers, bags, everything that can be used in the future. Also we use cloth towels and napkins instead of paper ones and use rechargeable batteries instead of disposable ones.

Try to minimize the usage of air conditioners, refrigerators and other electronic goods which cause hazardous effects to the mother earth and can maximize the usage of the solar energy and wind energy for the power generation. India is located between 8°-38°N and 68°-97°.5' E and exhibits extreme variation in altitude-from sea level to heights above vegetational limits in the Himalayas (ca 4500 m), and climate-from monsoon tropical in south to temperate and alpine in the north-western Himalayas and extremely arid to semi-arid in the north-western plains. It is floristically extremely rich with about 33 percent of its botanical wealth (over 15,000 species of higher plants) being endemic. There are about 141 endemic genera distributed over 47 families (Nayar, 1980). Further, of the 4,900 endemic species, larger percentage is localised in the Himalayas (about 2,532 species) than in other regions, namely, the peninsular tract (1,788 species), and the Andaman and Nicobar Islands (185 species). It is also estimated that floristic richness is maximum in the north-eastern region, which holds about 50 percent of India's total species diversity, *i.e.*, more than 7,000 species, and is considered as the cradle of flowering plants. Of 990 species of orchids, 700 species occur in this region (Nayar, 1989). This, thus, makes the Indian region botanically unique and interesting.

On the basis of distribution of flora, the Indian subcontinent has been divided into 8 to 10 regions. The sub-regions suggested by Chatterjee (1939) are: (1) Western Himalayas, (2) Eastern Himalayas, (3) Central Himalayas, (4) Assam, (5) Gangetic plain, (6) Indus plain, (7) Deccan, and (8) Malabar. Based on physiographic, climatic and cultural features, the ICAR recognised 8 agro-climatic regions (Murthy and Pandey, 1978), while recently, further micro-climatic considerations pin-point to 21 such regions. Also, the Planning Commission has demarcated the country into 15 agro-climatic zones. However, in the present account, broadly 8 regions have been recognised which more or less superimpose over the phyto-geographical regions of Chatterjee (1939). These plant/crop diversity regions, the zones proposed by the Planning Commission.

Enough evidence exists pointing to the range of crops of the Harappans, and the appearance of new crops later in the archaeological record, to date the evolutionary process in Indian crops within reasonable limits. The crops of the Harappan period were chiefly of West Asian origin. They included wheat, barley and peas. Of indigenous Indian origin were rice, tree cotton and probably sesame. Rice first appeared in Gujarat and Bihar, not in the centre of the Harappan culture in the Indus Valley. There is some rather doubtful evidence that African crops were also grown by the Harappans. There is a record of sorghum (*jowar*) from Sind and of *Pennisetum* (*bajra*) from Gujarat. Sesame recorded from Punjab is of uncertain origin, having been ascribed by some to Africa and by some to India.

The earliest record of the African cereal, *Eleusine coracana* (*ragi*) is from Mysore, about 1800 B.C. The South-East Asian crops of importance to India are sugarcane and banana, and they are not of the type likely to leave identifiable remains in the archaeological record. Consequently, nothing is known of them before historic times but both appear in the early literary record. Crops of American origin include maize, grain amaranths and potato. The dating of the introduction of maize is uncertain, the characteristics and distribution of some forms being such as to lend support to the view that they reached India in pre-Columbian times. The potato, on the other hand, was certainly a late introduction by either the Portuguese or the British.

The oldest crops have been under the selective pressures of the Indian environment for about 4500 years, though most of the crops except those of American origin have been exposed for at least 2000 years. The post-Columbian introductions from America, on the other hand, have been in India for less than 500 years.

It may also be emphasised here that crops of the Indian subcontinent have influenced the agricultural development of ancient Egyptian, Assyrian, Sumerian and the Hittite Civilization (Nile/Tigris/Euphrates river basins) through their early spread to these regions of the Old World. Also during the early Christian era, several Indian crops/plants were taken to South-East Asian countries by the Buddhists. Much early exchange of plant material had also existed with Africa, while many crops, *viz. Citrus*, cotton, jute, rice and sugarcane were distributed especially to the Mediterranean region by the Arabs in the 8-10th centuries A.D.

Similarly, several New World domesticates were introduced into India and reciprocally Indian cultigens were subsequently introduced into the New World (Mehra and Arora, 1982). Further, prevalence of great ethnic diversity and primitive agriculture (Arora, 1987) also points out to the rich heritage of the Indian subcontinent.

INDIAN SUBCONTINENT AS GENE CENTRE

India is one of the twelve centres/regions of diversity of crop plants in the world. As is evident from the above brief account, antiquity of agriculture as also the ethnic diversity in the sub-continent have played a major role in the diversification of crop resources in this region. Thus, rich genetic diversity occurs in several crop plants and their wild progenitors. As it is, the diverse agro/eco-climate of the 'Hindustani Centre', apart from possessing about 166 species of agri-horticultural crop plants, also has a rich diversity in wild relatives of crop plants numbering about 320 species. In addition, as already reflected above, the Indian agriculture has been enriched by a continuous stream of introductions of new crops and their cultivars by man since the ancient times. The Mughals, Spaniards and the Portuguese brought several new crops to this country.

Prominent amongst these were pear, grape, nut fruits, date palm, garlic, opium, maize, potato, sweet potato, tomato, chillies, French bean, peas, etc. The Arabs possibly brought clove, coriander, cumin, ammi or lavage (*ajwain*) and fennel. This process continued during the British period which saw the introduction of coffee, cocoa, cashew nut, strawberry, litchi, *Cinchona*, etc., besides collection of tea in Assam and adjoining region. The current matrix of diversity consists of the genepool of indigenous crop plants, their wild and/or weedy relatives and the well adapted introductions from practically all over the globe.

The introduced types, depending on the time of introduction, extent of material introduced and the areas of introduction, exhibit enormous build-up of diversity within the Indian subcontinent.

Some good examples are: *cereals*-wheat, barley, oats, maize; *pulses*-gram, French bean (*rajmah*) and peas; *vegetables*-potato, onion, cauliflower, cabbage, carrot and tomato; *fruits*-apple, pear, grapes, cherry, peach and apricot; *oilseeds*-soybean, sunflower and groundnut; *fibre crop*-cotton; to a very limited extent also in *medicinal plants*-mint, liquorice, *Digitalis* (foxglove), *Cinchona* (quinine), *Hyoscyamus* (henbane) and others such as *Humulus lupulus* (hops). Our knowledge on the origin of cultivated plants and regions of diversity has considerably increased since the time of de Candolle (1886) and Vavilov (1926). This has been amply elaborated.

By and large, these geographic areas vary from eight to twelve in number, lie between 20° and 45° latitude north and south, and are superimposed over the world's major vegetation formations/biomes/floristic zones (Good, 1953). The Indian gene centre is one such important region of diversity of crop plants. It has strong linkages and contiguity with the Indo-Chinese-Indonesian, Chinese-Japanese and the Central-Asian regions. Further, as pointed out above, the influx of germplasm in distant past from the Mediterranean, African and tropical American regions, has built up enormous locally adapted and selected variability.

It has already been stressed that the Indian subcontinent possesses rich floristic wealth. Of this 3,000-5,000 species are of economic value which include about 1,000 wild food plants. The diversity in plant genetic resources numbers around 166 species and the wild genepools of direct or indirect utility, around 320 species, distributed in eight phyto-geographical/agro-ecological zones.

By and large, India is:

1. a primary centre of diversity for crops, such as rice, black gram, moth bean, pigeonpea, cucurbits like smooth gourd, ridge gourd and pointed gourd, tree cotton, *capsularis* jute, jack fruit, banana, mango, *Syzygium cumini*/jamun, large cardamom, black pepper and several minor millets and medicinal plants like *Rauvolfia serpentina* and *Saussurea lappa.*
2. a secondary centre of diversity for African crops, such as finger millet, sorghum, cowpea, cluster bean, okra, sesame, niger and safflower; tropical American types like maize, tomato, pumpkin/*Cucurbita* spp., chayote or chou chou, chillies and *Amaranthus;* and with,
3. regional (Asiatic) diversity, for crops like maize, barley, amaranth, buckwheat, prosomillet, foxtail millet, mung bean/green gram, chickpea, cucumber, bitter gourd, bottle gourd, snake gourd and Brassicae.

Geographical contiguity with the Far-East and/or the Indo-Malayan (South/South-East Asian region) belt is largely responsible for more regional diversity in mung bean, rice bean, sword bean, tomato, citrus, small cardamom, *Saccharum*, ginger, turmeric, tuber crops, particularly taros and yams, and bamboos. Different agro-ecological/phyto-geographical regions of India, thus hold rich diversity in both the cultivated and the wild crop genepools. Based on the areas and crops, the following brief analysis on the distribution and extent of crop-plant diversity is given:

The following regions exhibit richness in crop plant diversity:

- *Western Himalayas including cold arid tract:* In Kinnaur, Lahaul and Spiti, Pangi, Ladakh and adjoining areas, rich genetic diversity occurs in wheat, barley (particularly hull-less types), buckwheat, prosomillet, amaranth, chenopods, field peas, lentil and several other crops, possessing adaptability to cold and tolerance to drought conditions. In barley, bluish/black grain types occur at high altitudes (above 3300 m), exhibiting akinness to local Tibetan types. In wheat, both awned and awnless, tall types occur and collections from Lahaul and Spiti have exhibited resistance/tolerance to rust. Further, rich diversity in rice, maize, French bean, peas, buckwheat, amaranth and other crops-radish, leafy Brassicae, pome, stone and nut fruits, forages and medicinal plants, occurs in the sub-humid hill regions of Jammu and Kashmir, Himachal Pradesh and Uttar Pradesh.
- *North-eastern region and the Eastern Himalayas:* The tribal dominated belts of Mizoram, Manipur, Meghalaya, Tripura, Sikkim, North Bengal and parts of Nagaland and Arunachal Pradesh, are rich in local variability of cereals-rice, maize (including the primitive pop-corn), barley, wheat; pseudocereals-buckwheat, *Chenopodium*, amaranth and

soft-shelled forms in *Coix;* millets-finger millet, foxtail millet; legumes-rice bean, winged bean, adzuki bean, black gram, sem/*Dolichos*, soybean, sword bean, peas; oilseeds-*Brassica* spp., *Perilla*, sesame, niger; vegetables-cucurbits (*Cyclanthera*, *Cucurbita*, *Momordica*, *Cucumis*, *Luffa*, *Lagenaria*, *Benincasa*)*;* fibres-*Corchorus* spp., tree cotton, kenaf and mesta; taro/yam and several other tuberous/ rhizomatous/bulbous types, and in *Citrus*, *Musa* and bamboos.

- *Eastern peninsular region:* The tribal dominated tract of Orissa and the Chotanagpur plateau may serve as good examples of areas that hold rich crop diversity *viz.* rice, sorghum, finger millet, fox-tail millet, prosomillet and other minor millets, *Dolichos*, rice bean, chickpea, pigeonpea, horsegram/kulthi, sesame, niger, Brassicae, linseed, castor, several cucurbits, brinjal, chillies, mango and wild types such as in rice (*Oryza nivara*).
- *Western arid/semi-arid region:* In Rajasthan and Gujarat (including Saurashtra), rich diversity occurs in sorghum, pearl millet, moth bean, cluster bean/*guar*, cowpea, black gram, green gram, Brassicae, sesame, cucurbits, forage legumes and grasses.
- *Central tribal region:* Madhya Pradesh and adjoining tract of Maharashtra possess rich diversity in wheat, rice, sorghum, minor millets, grain legumes-particularly cowpea, chickpea, pigeonpea, black gram and green gram; oilseeds-niger, sesame, Brassicae; and cucurbits.
- *Western peninsular region:* The Western Ghats tract has promising diversity in tuber crops-*Colocasia*, *Dioscorea* and *Coleus;* rhizomatous types like *Curcuma* species and ginger, and other crops like rice, finger millet, several grain legumes, and much variation in cucurbits, forage legumes and grasses; several fruit trees such as jack fruit and spices like cardamom and black pepper.

CROP DIVERSIFICATION IN THE INDIAN PERSPECTIVE

With the advent of modern agricultural technology, especially during the period of the Green Revolution in the late sixties and early seventies, there is a continuous surge for diversified agriculture in terms of crops, primarily on economic considerations. The crop pattern changes, however, are the outcome of the interactive effect of many factors which can be broadly categorized into the following five groups:

- Resource related factors covering irrigation, rainfall and soil fertility.
- Technology related factors covering not only seed, fertilizer, and water technologies but also those related to marketing, storage and processing.

- Household related factors covering food and fodder self-sufficiency requirement as well as investment capacity.
- Price related factors covering output and input prices as well as trade policies and other economic policies that affect these prices either directly or indirectly.
- Institutional and infrastructure related factors covering farm size and tenancy arrangements, research, extension and marketing systems and government regulatory policies.

Obviously, these factors are not watertight but inter-related. For instance, the adoption of crop technologies is influenced not only by resource related factors but also by institutional and infrastructure factors. Similarly, government policies-both supportive and regulatory in nature-affect both the input and output prices. Likewise, special government programmes also affect area allocation and crop composition.

More importantly, both the economic liberalization policies as well as the globalization process are also exerting strong pressures on the area allocation decision of farmers, essentially through their impact on the relative prices of inputs and outputs. Although the factors that influence the area allocation decision of farmers are all important, they obviously differ in terms of the relative importance both across farm groups and resource regions. While factors such as food and fodder self-sufficiency, farm size, and investment constraints are important in influencing the area allocation pattern among smaller farms, larger farmers with an ability to circumvent resources constraints usually go more by economic considerations based on relative crop prices than by other non-economic considerations. Similarly, economic factors play a relatively stronger role in influencing the crop pattern in areas with a better irrigation and infrastructure potential. In such areas, commercialization and market networks co-evolve to make the farmers more dynamic and highly responsive to economic impulses.

What is most notable is the change in the relative importance of these factors over time. From a very generalized perspective, Indian agriculture is increasingly getting influenced more and more by economic factors. This need not be surprising because irrigation expansion, infrastructure development, penetration of rural markets, development and spread of short duration and drought resistant crop technologies have all contributed to minimizing the role of non-economic factors in crop choice of even small farmers.

What is more, the reform initiatives undertaken in the context of the ongoing agricultural liberalization and globalization policies are also going to further strengthen the role of price related economic incentives in determining crop composition both at the micro and macro levels. Obviously, such a changing economic environment will also ensure that government price and trade policies will become still more powerful instruments for directing area allocation decisions of farmers, aligning thereby the crop pattern changes in line with the changing demand-supply conditions. In a condition where agricultural growth results more from productivity improvement than from area expansion, the increasing role that price related economic incentives play in crop choice can also pave the way for the next stage of agricultural evolution where growth originates more and more from value-added production. Turning now to the socio-economic and environmental consequences of crop pattern changes, the Green Revolution technologies have fomented, among other things, an increasing tendency towards crop specialization and commercialization of agriculture.

While these developments have positive effects on land/labour productivity and net farm income, they have also endangered a number of undesirable side effects like reduced farm employment and crop imbalances. Although the expansion of commercialized agriculture has fomented new sets of rural non-farm activities and strengthened the rural-urban growth linkages, it has also weakened the traditional inter-sectoral linkages between the crop and livestock sectors. Besides, crop pattern changes also lead to serious environmental consequences that take such forms as groundwater depletion, soil fertility loss and waterlogging and salinity-all of which can reduce the productive capacity and growth potential of agriculture over the long-term. A classical example is the rice-wheat system in Northwestern India replacing traditional crops like pulses, oilseeds and cotton.

The temporal analysis of the changes in crop pattern is observed both at the national and state level. For the purpose of the present study, crop pattern changes at the all India level are evaluated by considering the area share of crops and crop groups at four time points capturing, respectively, the triennium ending (The) averages of areas in 1966/67, 1976/77, 1986/87 and 1996/97. These time points have been selected so as to capture the major events and stages in the evolution of Indian agriculture which are of direct relevance to the purpose of this study.

The pre-Green Revolution situation is represented by the 1966/67 period. The post-Green Revolution situation is represented, on the other hand, by three sub-periods. These periods have been chosen deliberately to highlight the temporal differences in the factors contributing to agricultural growth. In the first of these three post-Green Revolution periods, *i.e.*, 1966/67 to 1976/77, the impact of the Green Revolution has mainly been on wheat, especially in the irrigation-wise better endowed Northwestern parts of India. But in the second period, *i.e.*, 1976/77 to 1986/87, the new technologies have expanded to other regions and also covered crops such as rice. During this period, the Southern and Eastern regions showed remarkable improvement in agricultural growth and productivity. More or less during the same period, the benefits of the new crop and irrigation technologies that were confined earlier to large and medium sized farms have also started percolating to the small farms. The third period, *i.e.* 1986/87 to 1996/97, is characterized by a number of policy changes including the launching of the Technology Mission on Oilseeds (TMO) as well as price support and stabilization policies for oilseed crops. During this period there is also a marked shift in the pattern of agricultural investment in favour of the private sector.

This period not only covers such important policy changes as unleashed by agricultural liberalization policies but also represents the consolidation phase of rural infrastructure. Since these four periods can capture well the changes in the set of factors that affect crop pattern, they can provide a better analytical framework for evaluating both the nature and technical context of crop pattern shifts observed not only at the national level but also at the state level. The analysis of crop pattern changes to be attempted at the macro level (national) will focus on three main aspects. These aspects are: a) the nature and direction of area shifts across crops and crop groups observed through time, b) the implications of these shifts for crop diversification and balance in the inter-crop allocation of existing and additional areas brought under cultivation, and c) the output and productivity impact of crop pattern changes.

CROP DIVERSIFICATION IN INDIA

India is a country of about one billion people. More than 70 percent of India's population lives in rural areas where the main occupation is agriculture. Indian agriculture is characterized by small farm holdings. The average farm size is only 1.57 hectares. Around 93 percent of farmers have land holdings smaller than 4 ha and they cultivate nearly 55 percent of the arable land. On the other hand, only 1.6 of the farmers have operational land holdings above 10 ha and they utilize 17.4 percent of the total cultivated land. Due to diverse agro-climatic conditions in the country, a large number of agricultural items are produced. Broadly, these can be classified into two groups-foodgrains crops and commercial crops.

Due to the challenge of feeding our vast population and the experience of food shortages in the pre-independence era, 'self reliance' in foodgrains has been the cornerstone of our policies in the last 50 years. Around 66 percent of the total cultivated area is under foodgrain crops (cereals and pulses). Concurrently, commercial agriculture developed for whatever reasons in the pre-independent phase also kept flourishing during the post independent period. Commercial agriculture not only catered to the domestic market but has also been one of the major earners of foreign exchange for the country.

Crop diversification is intended to give a wider choice in the production of a variety of crops in a given area so as to expand production related activities on various crops and also to lessen risk. Crop diversification in India is generally viewed as a shift from traditionally grown less remunerative crops to more remunerative crops.

The crop shift (diversification) also takes place due to governmental policies and thrust on some crops over a given time, for example creation of the Technology Mission on Oilseeds (TMO) to give thrust on oilseeds production as a national need for the country's requirement for less dependency on imports. Market infrastructure development and certain other price related supports also induce crop shift. Often low volume high-value crops like spices also aid in crop diversification. Higher profitability and also the resilience/stability in production also induce crop diversification, for example sugar cane replacing rice and wheat. Crop diversification and also the growing of large number of crops are practiced in rainfed lands to reduce the risk factor of crop failures due to drought or less rains. Crop substitution and shift are also taking place in the areas with distinct soil problems. For example, the growing of rice in high water table areas replacing oilseeds, pulses and cotton; promotion of soybean in place of sorghum in vertisols (medium and deep black soils) etc.

TEMPORAL CHANGES IN THE AREA SHARE OF CROPS

The temporal behaviour of crop pattern changes at the all India leve, respectively, the area share of main crop groups and major crops for the four periods. Though obvious, it needs to be stated that the changing area share of crops is due as much to shift in area under other competing or alternative crops as to the relative area allocation of fresh areas brought under cultivation. In any case, the changing area share of crops does capture the ongoing changes in the comparative advantage calculus of farmers. The changes in the comparative advantage of crops reflect, in reality, the ongoing changes in relative prices of inputs and outputs, production conditions (including irrigation expansion), development and spread of new crop and farm technologies, extension and input support policies and trade policies and domestic regulations. As such, the changing area share of crop pattern, though looking deceptively simple, becomes a useful tool for understanding the direction in which crop pattern changes are influenced by the variations in the comparative advantage of crops and crop groups.

Although cereals gained a marginal increase in area share in the first decade of the Green Revolution, their area and share declined gradually thereafter. Between 1966/67 and 1996/97, 3.35 percent of the gross cultivated area (GCA)-representing approximately about 5.7 million hectares (m/ha)-has shifted from cereal crops to non-cereal crops. Since the area share of pulses taken as a group also declined by 1.57 percent during the same period, the area share of foodgrains as a group declined by 4.92 percent during 1966-97. In area terms, the shift from foodgrains to non-foodgrains involves an approximate area of about 8.36 m/ha. While cereals and pulses have lost area, the major gainers of this area shift are the non-foodgrain crops especially oilseeds. The area share of oilseeds as a group that has gone up by 4.08 percent accounts for about 83 percent of the 8.36 m/ha involved in the area shift between 1966/67 and 1996/97. As we consider the share of individual crops within cereals, although the share of cereals as a group has declined, the area share of rice has increased continuously over all the four periods.

Wheat, although having a declining area share until 1986/87, also gained in its share when the entire period is considered. Thus, the area loss of cereals can be attributed entirely to the declining area share of coarse cereals, especially sorghum, pearl millet, barely and small millets. It can be noted that even within coarse cereals, the area share of maize shows a marginal improvement over the years. Within oilseeds, the crops showing steady improvement in their area share are: rapeseed and mustard, soybean and sunflower. Among these three oilseeds gaining in area share, rapeseed and mustard are substantially grown as intercrops with wheat.

On the other hand, the area shares of other oilseeds including groundnut (that has a dominant area share within oilseeds) but excluding coconut, which is more a plantation crop than field crop, have either fluctuated or declined. The area share of groundnut, though improved during the last period, has declined as compared to its share in the pre-Green Revolution period. But, the declining area share of crops-especially those with only a marginal change in their area share-need not necessarily imply a decline in the actual area under these crops. Since the Gross Cropped Area (GCA) is constantly increasing over time, partly through an expansion of net sown areas as in the initial stages of the Green Revolution and partly through increasing intensity of cropping mainly by irrigation expansion, the declining area share can coincide with an increase in absolute increase in the area under crops.

Although the increase in the area share of other commercial crops is not as dramatic as that of oilseeds, it is still notable because of its implications for the direction of Indian agriculture. But, among these other commercial crops that cover fibres, spices, fruits and vegetables, and other field crops such as tobacco and sugar cane and plantation crops, only spices, fruits and vegetables show a steady improvement in their area shares, whereas others show mostly a declining trend. This is particularly true for fibres and other field crops that have over four fifths of the total area under the broad group of other commercial crops. However, sugar cane, included in the category of other field crops, shows an increase in its area share. This is also true for cotton included in the fibre category. While all spice crops show a gradual increase in their area share, only three of the six crops included in the fruits and vegetables category show a gain in their area share over the years. These crops are banana, potato and onion.

It is clear that most of the area shifts that occurred during the three decades between 1966/67 and 1996/97 is from coarse cereals to oilseeds. Three notable aspects of this area shift towards oilseeds can be noted. First, most of these area shifts have occurred particularly during the decade ending 1996/97. A protective trade environment, favourable price policy and the connecting of the Technology Mission on Oilseeds (TMO) during this period have all enhanced the comparative advantage of oilseeds.The area share of oilseeds that increased just by a percentage each during the two decades before 1986/87 has risen suddenly by 43 percent during the decade ending 1996/97. Second, the area shift during this period comes mainly from barley and pulses other than pigeon pea.

Since most of those crops losing their area share are usually grown under rainfed conditions where oilseeds can also be grown, the area shift can be said to involve mostly rainfed areas, although comparative advantage and crop rotation considerations often favour oilseeds even in groundwater irrigated areas. There is also a significant area shift within oilseed crops. For instance, while the area shares of rapeseed and mustard, sunflower and soybean are increasing steadily, those of sesamum, linseed and nigerseed are declining gradually. Thus, the area shift has favoured only a sub-sector within the oilseed sector partly because of constant changes in the comparative advantage of different oilseeds and partly because of the impact of changing consumers' preferences on the relative demand of oilseeds.

As noted already, area shifts and crop pattern changes can lead either to crop specialization or to crop diversification. The area share of foodgrains increased during 1967-76 due partly to their yield advantages created by irrigation expansion and Green Revolution technologies and partly to government policies pursued to encourage food production and eliminate food imports. As a result, there was a tendency towards cereal centered specialization.

But, later when increased productivity of foodgrains, especially cereals, made it possible to allocate more area to other crops such as oilseeds with a severe supply shortage, the specialization tendency witnessed earlier has given room for overall crop diversification. Even within such an overall crop diversification, it is also possible to see specialization tendencies within each crop group. For instance, within cereals, the declining share of coarse cereals and increasing share of wheat and rice indicates a rice and wheat centered specialization. Such a specialization has become possible because within the achievement of self-sufficiency, mainly through an expanding production of wheat and rice, the food implication of coarse cereals has declined almost coinciding with their declining demand prompted by an increasingly incurring income level. On the other hand, within oilseeds, even though groundnut still has a dominant area share, the growth of area under rapeseed and mustard, sunflower and soybean indicates an ongoing structural change leading to diversification within the oilseed sector. The issue of crop diversification, although considered in area terms, assumes significance in view of its effects on the supply-demand balance of main crops and crop groups. The rice and wheat-centered specialization, for instance, indicates an increase in the supply of wheat and rice but a reduced supply of coarse cereals. Since the demand for coarse cereals is declining and that for wheat and rice is increasing due to changes in the income pattern, the changes in their supply are actually necessary to achieve the required demand-supply balance. A similar line of argument can also be extended to other crops where in addition to domestic demand, international demand and supply also assume significance.

Clearly, during the Green Revolution period, the foodgrains category has evinced an increasing tendency towards specialization with an unbalanced area composition, whereas the non-foodgrains category has shown an increasing tendency to diversify with an improved balance in their area composition. But, during 1976-97 the non-foodgrains category has also caught up with the specialization tendency and area imbalance that characterized foodgrains throughout the entire period. The reason for the specialization tendency within foodgrains is rather obvious in view of the declining area share of coarse cereals and the increasing area share of rice and wheat.

The changing area share of crops can also explain the specialization tendencies observed among the non-foodgrains. As we have seen in the context, the area shares of oilseeds and other commercial crops have increased especially during the post Green Revolution period characterized by a prices and market-responsive agriculture inevitable in the aftermath of achieving food self-sufficiency. It is important to note, however, that oilseeds are more diversified with an enhanced balance in their area composition.

Within foodgrains, on the other hand, while cereals tend towards specialization with an increasing imbalance in their area composition, pulses evince a tendency to diversify with a declining imbalance in their area composition. The latter result is mainly due to the fact that the area shares of the dominant pulses such as gram have declined to give room for other pulses such as pigeon pea. But it should be noted that diversification and area balance achieved by pulses have occurred only with a reduced area share. Among the non-foodgrain crops, spices are notable not only for their increasing tendency towards diversification but also for a declining imbalance in their area composition. In contrast, fibres, fruits and vegetables, other field crops and plantation crops have all evinced a specialization tendency especially after 1986/87.

5

Revisiting the Fundamentals of Competitiveness

FOOD INDUSTRY

Since the late 1970s, the food industry has emphasized an organized sanitation programme that monitors the microbiology of raw ingredients in production plants and the wholesomeness and safety of the finished products, in an effort to maintain or upgrade the acceptability of its food products.

As consumers become better informed and more sophisticated, it is even more vital for the food industry to develop an effective quality assurance (QA) and sanitation programme. The efforts of regulatory agencies in the field of sanitation and food microbiology have been responsible for the food industry's implementation of voluntary quality assurance programmes. Food scientists have also had a positive impact on quality assurance programmes because many of these professionals have joined various companies in the food industry. Their efforts have been instrumental in the adoption and/or upgrading of quality assurance programmes for the organizations they represent.

In its initial stages, quality assurance was primarily a quality control function, acting as an arm of manufacturing. It has now evolved to formidable force within executive structure of large food firms and has emerged into broad spectrum of activities. A quality assurance programme provides the avenue to establish checks and balances in the areas of food safety, public health, technical expertise, and legal matters affecting food manufacturing firms. Activities related to food sanitation include sanitation inspections, products releases and holds, packaging sanitation, and product recalls and withdrawals.

A quality assurance programme that emphasizes sanitation is vital to the growth of a food establishment. If foods are to compete effectively in the market place, established hygienic standards must be strictly maintained. However, it is sometimes impractical for production personnel to measure and monitor sanitation while maintaining a high level of productivity and efficiency. Thus, an effective quality assurance programme should be available to monitor, within established priorities, each phase of the operation. All personnel should incorporate the team concept to attain established sanitary standard, ensuring that food products in the market place are safe.

INCREASE THE EFFECTIVE USE OF AGRICULTURAL BIODIVERSITY IN FOOD AND FIBRE PRODUCTION

RATIONALE

Agricultural biodiversity performs vital functions in agriculture, land and water use. The diversity of plants, animals and microorganisms is essential for maintaining the productivity and sustainability of farm crops and animals, managed forests and rangelands, aquaculture and fisheries. Future global food security is dependent on harnessing and sustaining agricultural biodiversity and its many functions, from the farm plot to the landscape level. In both low external input and high input agriculture, the goals of sustainability, productivity and equity may best be met through agro-ecosystem designs that enhance functional diversity at the genetic, species and landscape levels. A central challenge across the whole range of agroecosystems is to find alternatives to the input substitution approach and future dependence on costly biotechnology packages.

This can be achieved through an agro-ecological approach that seeks to break the monoculture structure and dependence on suppliers of off farm inputs through the design of integrated agroecosystems.

By assembling a functional biological diversity within and around agroecosystems, it is possible to encourage synergism's that subsidise agro-ecosystem processes by providing ecological services, the recycling of nutrients and the enhancement of natural enemies of pests as well as provide diverse, quality foods and other farm products.

The current emphasis on genetic engineering must be balanced by higher level approaches that build on agro-ecology, landscape ecology as well as social and biological diversity. National sovereignty and food security ultimately depend on a wide choice of agricultural technologies and development options.

ACTIONS

- Establish and implement national policies to support agro-ecological approaches that enhance agricultural biodiversity in food and fibre production through integrated pest, crop, nutrient and soil management as well as land use planning. The policy framework should ensure political commitment, educational and institutional capacities as well as incentives for enhancing agricultural biodiversity and its functions in agriculture and land use

- Provide fiscal and administrative support for studies that investigate ways to maintain and enhance the utilisation of agricultural biodiversity in crop and animal production and in different kinds of agroecosystems, particularly high yielding and intensive commercial production systems
- Promote development on the basis of locally adapted genetic material. Strengthen in particular the capacity to develop new crop varieties and animal breeds that are specifically adapted to local environments. Increase the range of genetic diversity available to farmers
- Broaden the use of genetic diversity to protect crops and animal breeds against pest and weather problems by introducing multiple genetic systems for coping with stresses and also by deploying functional genetic mixtures and multilines where appropriate
- Broaden the use of species diversity into functional designs for agroecosystems that sponsor more of their own soil fertility, crop protection, pollination and water management (agroforestry systems, multiple cropping, fish polycultures...).
- Plan and manage rural landscapes to sustain biological diversity and ecosystem services. Where appropriate, maintain hedgerows, windbreaks and mangrove strips, leave tracts of land in native habitat, plant a diversity of crops, encourage pastoral activities, mixed species forestry and aquaculture. Maximise the use of resources internal to the landscape whilst closing nutrient cycles by integrating production with local needs and markets situated within or near the managed landscape. Apply the same principles for urban landscapes where food and fibre are produced.

PROMOTE LOCAL ADAPTIVE MANAGEMENT OF AGRICULTURAL BIODIVERSITY

RATIONALE

Variation within and among agroecosystems is enormous. Daily, seasonal and longer term changes in the spatial structure of agricultural biodiversity are apparent at the broad landscape level right down to small plots of cultivated land. These spatio-temporal dynamics have major implications for the way agricultural biodiversity is managed,-how, by whom and for what purpose.

Uncertainty, spatial variability and complex non-equilibrium and non linear ecological dynamics emphasise the need for flexible responses, mobility and local level adaptive resource management in which local users of agricultural biodiversity are central actors in analysis, planning, negotiations and action. This calls for far greater appreciation of local farming practices and knowledge used by rural people to manage agricultural biodiversity in forests, wetlands, fields, rangelands, coastal zones and freshwater systems. It frequently suggest new practical avenues for technical support in which land users' own priorities, knowledge, perspectives, institutions, practices and indicators gain validity.

VARIATION IN AGRICULTURAL BIODIVERSITY

- Crop varieties planted out experience rapid changes in environmental conditions, both above and below ground. For example, the physico-chemical and biological characteristics of soils are rarely homogenous within a single plot, let alone between plots. The intense selective pressures associated with this kind of microgeographical variation calls for a fine grained approach to agricultural biodiversity management that hinges on participatory plant breeding and decentralised seed multiplication. This adaptive strategy is generally advocated for resource poor farming systems in marginal, risk prone environments. However, it may be increasingly relevant for high input situations where agricultural diversification is used to solve production problems induced by genetic uniformity (*e.g.* pest outbreaks) or to exploit new market opportunities (economic niche for local or regional products).
- The abundance of insect pests and their predators is known to vary enormously within and between fields, even in the more intensively managed systems. In high input irrigated rice farms, 100 fold differences in abundance of planthopper populations are commonly observed on rice plants grown a few metres apart. Huge variations in insect abundance also exist at larger spatial scales and all are marked by dynamic change over time. This implies that highly differentiated pest management approaches are needed to monitor and control pests effectively and economically. The FAO-Government programme on IPM is a clear demonstration of the advantages of such local adaptive management of pests and their predators in irrigated rice in Asia. Farmer Field Schools have been a major innovation for the local adaptive management of agricultural biodiversity by developing the capacity of farmers to develop site-specific crop protection solutions.

ACTIONS

- Carry out administrative tasks, land use planning, agricultural research and development as near to the level of actual users of resources or beneficiaries of administration as is compatible with efficiency and accountability

- Ensure flexibility and diversity in institutional and organisational design to enable government administration and services to track appropriately the dynamic changes which occur in agroecosystems and the functions of agricultural biodiversity at different time and spatial scales.
- Educate policy makers, professionals and the public (including the bearers of local knowledge) about the value of local and indigenous knowledge and management systems in sustaining agricultural biodiversity and its many functions
- Strengthen local groups and institutions by devolving resources and removing administrative or legal hurdles to local planning and action. Support the development of local institutions for common property resources and the equitable sharing of benefits from their utilisation
- Identify and support a mediator for conflict resolution and an arbiter of last resort; guaranteeing a level legal playing field and equality of advocacy in disputes, both within and between local groups as well as between local groups and powerful external interests. Of particular importance in this connection are government policies and actions that explicitly prevent discrimination on the basis of differences in gender, ethnic origin and wealth.

SUPPORT LOCAL PARTICIPATION IN PLANNING, MANAGEMENT AND EVALUATION

RATIONALE

From the outset, the definition of *what* agricultural biodiversity is to be conserved, *how* it should be managed and *for whom* should be based on interactive dialogue to understand how local livelihoods are constructed and people's own definitions of well being. Most professionals have tended to project their own categories and priorities onto local people and landscape management. In particular, their views of the realities of the poor, and what should be done, have generally been constructed from a distance. Household livelihood strategies often involve different members in diverse activities and sources of support at different times of the year. Many of these, like collecting wild foods and medicine, home gardening, common property resources, share-rearing livestock and stinting are largely unseen by outside professionals.

Agricultural Research and Development and land use planning should start with enabling local people, especially the poor, to conduct their own analysis and define their own priorities. This methodological orientation is absolutely essential to meet the goals of equity, sustainability, productivity and accountability. In that context, dialogue, negotiation, bargaining, conflict resolution and joint management agreements are all integral parts of a long term participatory process which continues well after the initial appraisal and planning phases into monitoring and evaluation. This implies the adoption of a learning process approach in the management of agricultural biodiversity and its functions. It also calls for a new professionalism with new concepts, values, participatory methodologies and behaviour.

ACTIONS

- Ensure public participation of women and men (particularly farmer, herder, fisherfolk and forest dweller involvement) in the development of land use and agricultural policies and in the generation of technologies
- Ensure inclusive equitable representation (gender, class, ethnic origin, age) in the participatory activities and process
- Provide capacity building for technical and scientific personnel to foster those participatory skills, attitudes and behaviour needed to learn from farmers and rural people (mutual listening, respect, gender sensitivity as well as methods for participatory learning and action)
- Provide institutional space and incentives for professionals to understand social and cultural complexity as well as agro-ecological diversity
- Support joint problem-solving among local people, scientists and/or extension workers and the development of negotiated participatory research agendas and resource management agreements, using local criteria and indicators as well as those of outside professionals and their organisations
- Support the participatory monitoring and evaluation of national policies, land use plans, food and fibre production technologies by building on the perspectives of all social actors. Encourage the use of gender desegregated and socially differentiated local indicators and criteria in monitoring and evaluation as well as in guiding subsequent technical support, policy changes and allocation of scarce resources for agricultural biodiversity management.

TRANSFORM BUREAUCRACIES AND PROFESSIONAL PRACTICE

RATIONALE

Local adaptive management of agricultural biodiversity and large-scale participation do not mean that state bureaucracies and other external organisations have no role. But they do challenge bureaucracies to assume different roles and responsibilities. In particular, existing bureaucracies and professionals will often need to shift from being project implementers to new roles that facilitate local people's analysis, planning, action, monitoring and evaluation. The whole process should lead to local institution building or strengthening, so enhancing the capacity of people to take action on their own. Appropriate partnerships and co-management agreements between states, the private sector and rural communities are also required through new legislation, policies, institutional linkages and processes.

However, the adoption of a participatory culture and changes in professional attitudes and behaviour are unlikely to automatically follow when new methods are adopted. Training of agency personnel in participatory principles, concepts and methods must be viewed as part of a larger process of reorienting institutional policies, organisational cultures, procedures, financial management practices, reporting systems, supervisory methods, reward systems and norms. In both government departments and non governmental organisations, the challenge for top and middle management is to design appropriate institutional mechanisms and rewards to encourage the spread of participatory methods within the organisation.

Without this support from the top, it is unlikely that participatory approaches that enhance local capacities and innovation will become core professional activities. An important challenge concerns the creation of new relationships between local communities and their representative organisations, government agencies, research institutions and transnational corporations. Capitalising on the energy, resource efficiency and dynamism of the private sector constitutes a significant opportunity for increasing the performance of the agricultural sector and environmental sustainability.

ACTIONS

- Encourage shifts from hierarchical and rigidly bureaucratic structures to "flat", flexible and responsive organisations
- Ensure that senior and middle management positions are occupied by competent facilitators of organisational change, with the vision, commitment and ability to reverse gender and other discriminatory biases in the ideologies, disciplines and practices animating an organisation.
- Promote and reward management that is consultative and participatory rather than verticalist and efficiency led. Establish incentive and accountability systems that are equitable for women and men
- Provide incentives and high rewards for staff to experiment, take initiatives and acknowledge errors as a way of learning by doing and engaging with the diverse local realities of farming, fishing and pastoral societies

- Redesign practical arrangements, the use of space and time within the workplace to meet the diverse needs of women, men and older staff as well as their new professional obligations to work more closely with farmers and other stakeholders (time tables, career paths, working hours, provision of paternity and maternity leave, childcare provisions, mini sabbaticals, promotion criteria...)
- Support the creation of mechanisms of communication and collaboration between the private sector, and in particular, major corporations active in the agricultural sector, and other kinds of institutions and groups

AGRICULTURAL BIODIVERSITY'S CONTRIBUTIONS TO PRODUCTION AND ENVIRONMENTAL SUSTAINABILITY

Each species in an agro-ecosystem is part of a web of ecological relationships connected by flows of energy and materials. Whilst each species may occupy a specific ecological niche (*e.g.* primary producer, specialist or generalist consumer, decomposer) it is involved in sustaining many different agro-ecosystem functions and environmental processes, either directly or indirectly. In this sense the different components of agricultural biodiversity are inherently multifunctional and contribute to the resilience of production systems whilst providing environmental services at the larger landscape level. However, it is important to note that some species may play a key driving role in forming the structure and overall behaviour of agro-ecosystems and landscapes at different scales. There is indeed growing evidence that the diversity and functional complexity of all ecosystems can be traced to a small number of critical structuring processes, some of which are mediated by critical "keystone species". An example is the suite of 35 species of insectivorous birds that mediate budworm outbreak dynamics in the eastern boreal forest of North America.

Farmers, herders and fishermen have often enhanced the multiple functions of agricultural biodiversity through choice of genetic material, design of cropping patterns, development of crop and livestock production systems, land and water management practices as well as institutional arrangements. Local knowledge about the properties and dynamic roles of agricultural biodiversity is crucial in this connection. For example, the *mal monte* (bad weeds) and *buen monte* (good weeds) management systems of Mexican farmers recognises that the vegetation community as a whole must be managed to promote those aspects that are beneficial.

In low external input farming, different components of agricultural biodiversity are usually combined to give practices finely tuned to the local biophysical and socioeconomic conditions of individual farmers, herders and fish culturists. Natural processes mediated by agricultural biodiversity are favoured over external inputs and by products or wastes from one component of the agro-ecosystem become inputs to another. An example is the mulberry grove-fishpond system in the Pearl River Delta of China. In this multifunctional system, the white mulberry (*Morus alba*) tree produces organic substances (mulberry leaves etc).

These are used to feed silkworms that, in turn, produce their silk and chrysalides. The fallen parts of the mulberry tree and the excrement of the silkworm are applied to the fishpond where they are converted into fish biomass. The excrement of the fish, as well as other unused organic matter and bottom mud are returned to the mulberry grove as fertiliser, after being broken by a diverse suite of benthic micro-organisms. The agricultural biodiversity harnessed by the fish culturalists allows for the closing of nutrient cycles and efficient production in time and space. Fish polycultures are thus made up of species that dwell in the upper, medium and lower layers of the pond, as well a fish species with different feeding habits (*e.g.* plankton feeders, herbivorous fish, benthic mollusc feeders, and omnivorous fish).

In more specialised, high input farming based on the use of high yielding varieties, agricultural biodiversity helps sustain many production functions such as soil organic matter decomposition, pollination and pest control. In the USA or Australia for example, farmers may manage cover crops primarily to save soil and water in intensive orchard production systems. However, the species chosen will usually perform other functions in the agro-ecosystem. In addition to protecting against soil erosion, cover crops usually enhance soil structure, improve soil fertility and nutrient cycling as well as play a role in pest management by providing habitat heterogeneity and preserving a favourable balance between pests and predators. Depending on the species, trees can also provide fodder for animals, so increasing the number of internal linkages within the agro-ecosystem. These examples highlight the multiple functions of agricultural biodiversity and are intended as a reminder that functions are discussed one by one in this report only for the sake of convenience.

Decomposition and nutrient cycling functions. The crop plants, trees, livestock and fish deliberately chosen by farmers are the main determinants of the diversity of the flora and fauna that makes up the decomposer subsystem (Swift and Anderson, 1993). Available evidence shows that decomposer communities are highly diverse and are centrally involved in nutrient cycling, organic matter dynamics and other ecosystem functions. Detailed knowledge on the extent and functions of this diversity is limited; there is relatively more information on the functions of soil biodiversity than on the dynamics of decomposer communities in aquatic environments. Some functional groups in soils are widespread in distribution (*e.g.* nitrogen fixing bacteria, mycorrhizal fungi and predators of soil borne pests) whilst other like earthworms and termites are more restricted in their distribution. A gradually emerging picture structures soil biodiversity into a series of more or less spatially independent guilds. The spatial separation between distinct guilds (surface litter, root litter, rhizosphere guilds...) allows decomposer organisms to co-exist whilst containing communities of organisms that are functionally equivalent. For instance, several ecological guilds of earthworms can be recognised in humid tropical soils with different roles in litter transformation and as "ecosystem engineers" (Lavelle et al 1998). Through their activities of feeding, burrowing and casting, they modify the physical, chemical and biological properties of soil and thus its ability to support above ground vegetation. Together with termites, different species of earthworms are among the key functional groups in humid tropical soils. At least 42 native and exotic earthworm species common in tropical agro-ecosystems have been identified as able to resist disturbances linked to agriculture and agroforestry practices and build up sizeable populations in these environments.

The sensitivity of decomposer functions to farm management practices is also evident in the mechanised agriculture of developed countries. Comparisons of the soil biological diversity in biodynamic, organic and conventional farms in Switzerland show higher species diversity and functional levels in biodynamic and organic plots than in conventional systems. The significantly higher biomass, diversity and functional activity of soil micro-organisms, earthworms, ground beetles, staphilinids and spiders found in biological systems are largely due to the organic amendments and more selective plant protection measures used in the biological systems.

In high input-high output agriculture, microbial diversity is also a central component of integrated plant nutrition systems (IPNS) that aim to maximise the efficiency of plant nutrient supplies to crops by complementing the use of on-and off-farm sources of plant nutrients. Nitrogen fixation through bacteria (notably *Rhizobia spp*) and algae (*Azolla spp*) as well as phosphorus cycling *via* mycorrhizal fungi species are particularly noteworthy in this connection.

Microbial diversity is generally known to mediate nutrient cycling. However, there are few detailed studies of the dynamic role of micro-organisms in structuring landscapes and agro-ecosystems at different spatial and temporal scales. A long term study of African savannahs has shown that the productivity of large mammalian herbivores,-upon which the human economy of the savannah is based, is dependent on water availability acting as an "on-off" switch for the mineralisation of nitrogen, phosphorus and sulphur. Some 80% of the mineralisation is performed by the diverse soil micro-organisms which respond dramatically to the presence of water or rain. Where the immediate limiting resource in broad leafed savannahs is nitrogen, the key effect of water is to control the availability of inorganic nitrogen by modulating the functions of soil microbial diversity rather than control photosynthesis.

The interaction between water and micro-organism activity thus sustains soil fertility. In turn, soil fertility has a profound effect on savannah ecology by determining not only plant production but also what fraction of it is edible and which species of plants and animals will be present. Livestock production is thus closely dependent on the dynamic interactions between water availability and the activity of a diverse suite of soil micro-organisms in these semi-arid landscapes.

Biomass production and yield efficiency functions. Low external input production systems usually incorporate a wide range of species and genotypes that serve a variety of production goals and are used for their resistance to diseases and pests as well as for the differential exploitation of microhabitats. The relative productivity and efficiency of these diverse agro-ecosystems (fish polycultures, mixed herds, intercrops, integrated agro-silvo-pastoral systems) have been quantified in terms of relative yield or energy efficiency of diverse units as compared with sole crops. Results indicate that diversity rich agriculture is generally highly productive in terms of its use of energy and unit land area (or unit water volume).

For example, the land equivalent ratio (LER) of yields with intercrops in which mixtures include a legume species is usually significantly greater than outyield of sole crops. The energy efficiency (ratio of energy output to energy input) of pig or poultry production in internally diverse agro-ecosystems can be up to 10 times higher than that of intensive pig and poultry farms based on genetically uniform single species reared with enormous subsidies of fossil fuels. Whilst the yield output per labour hour of the more intensive and uniform systems is extremely high (in the absence of the internalisation of social and environmental costs), the more agricultural biodiversity rich systems are generally efficient producers of significant amounts of biomass. This efficiency is largely a product of the systems' biological and structural complexity that increases the variety of functional linkages and synergies between different components of agricultural biodiversity.

Soil and water conservation functions. Soil, water and nutrient conservation have been improved with the use windbreaks, contour farming with appropriate border crops and cover crops in a wide range of agro-ecosystems. In France over 150 species of trees and shrubs are used for soil and moisture conservation, with different species mixtures planted as hedges and taller windbreaks in gardens, orchards, whole farms and the larger rural landscape. In Sahelian countries of Africa windbreaks made up of *Euphorbia tirucali*, *Parkinsonia aculeata*, *Opuntia tuna* and *Prosopis africana* trees and shrubs help to conserve soil and moisture, and raise the yields of cereals which are grown between. In Mexico, contour lines are often planted with *Agave americana* to conserve soils and retain moisture whilst in southern Italy and Greece *Opuntia tuna* performs similar functions.

Many of these plants are multiple purpose species yielding wood, edible fruit and nuts, fodder, refuges for natural enemies of pests, nitrogen biofixation and medicines in addition to their soil and water conservation functions. Cover crops consist of plant species that are deliberately established after or intercropped with a main crop to serve various regenerative and conservation functions including soil and water conservation. Annual bluegrass, lana vetch, crimson clover, black medic, purple vetch and barley are some of the species recommended as cover crops for orchards and vineyards in California, USA. The wide variety of management systems in high input-high output orchards and vineyards creates a demand for a diversity of cover crops. Grass species have fibrous root systems that make them particularly useful in building soil structure, providing erosion control, and improving water penetration. Legumes are not as effective as grasses in improving water penetration but they contribute nitrogen to the soil. Many cover crop options can be selected for soil and water conservation from a diversity of annually seeded winter growing grasses and legumes, summer annuals, perennial grasses and legumes and reseeding winter annual grasses and legumes.

In North America and Europe, living mulch systems can be an economic way for commercial soybean, corn and vegetable growers to reduce soil erosion and water loss, increase soil organic matter and keep yields constant in high input-high output agroecosystems. Legume species commonly used as living mulches include alfalfa, short white clover, hairy vetch and red clover.

Pest control functions. Agricultural biodiversity in the form of predators, parasitic wasps, microorganisms plays a key role in controlling agricultural pests and diseases. For example, according to CAST (1999) more than 90% of potential crop insect pests are controlled by natural enemies that live in natural and semi-natural areas adjacent to farmlands. They have estimated the substitution of pesticides for natural pest control services at a cost of $54 billion per year.

Many methods of pest control,-both traditional and modern-, rely on biological diversity. The development of crop varieties and animal breeds that are resistant to specific pests and diseases selectively draws on the genetic diversity available *in situ* and in *ex situ* collections of germplasm. Genetic mixtures deployed in temperate and tropical agroecosystems can be effective in containing diseases in small grain crops as well as insect outbreaks in cassava, corn and potato for example. There are also many documented experiences showing that insect pests tend to be less abundant and damaging in agroecosystems with higher plant diversity *e.g.* intercrops, polycultures, crop rotations, cover crops, mixed tree stands, mixtures of annual and perennial plants. Depending on the pest species and the context, the plant diversity acts to reduce pest damage by interfering with the host seeking and reproductive behaviour of the pest, by enhancing the pests' natural enemy populations or by a combination of these processes. Judicious vegetation management within and around agroecosystems can thus enhance biological control or confer an overall resistance to pests and disease outbreaks.

Understanding how agricultural biodiversity directly or indirectly affects pest and disease dynamics is critical for the design of pest management at different scales. For instance, recent work in Javanese rice fields shows that there is an enormous diversity of arthropods, even in high input-high output agriculture. The arthropod communities are structured such that the dynamics of seasonal succession consistently lead to high levels of pest suppression, with little chance of outbreak. From the time that water first floods a farmer's field in preparation for planting, organic matter,-derived from residues from the previous crop cycle, organic waste in irrigation water, and algal growth-, provides the energy for an array of microorganisms (bacteria and phytoplankton) and detritus-eating insects.

The adults of the plankton-feeders (midges and mosquitoes) together with the detritus-feeders, provide a consistent and abundant source of alternative food for generalist predators very early in the season. As a result, pest mortality due to predation is high from the very earliest part of the season; hence, minimising the chance of damaging pest outbreaks. However, this intrinsic strength and stability of the rice agro-ecosystem is influenced by two main, large scale, external factors: 1) local and regional patterns of pesticide use, and 2) landscape effects-specifically, the spatial scale at which fields are synchronously planted, the duration and nature of fallow periods, degree of surrounding weedy or natural vegetation and existence of nearby ponds or other sanctuaries for natural enemies. This type of information on the functions of agricultural biodiversity in rice paddies provides the ecological basis for integrated pest control through careful management of the wider landscape and decisions on pesticide use.

Pollination and dispersal functions. There are more than 100,000 known pollinators (bees, butterflies, beetles, birds, flies, and bats). Pollination mediated by components of agricultural biodiversity is an important function in a variety of terrestrial agroecosystems (biotic pollination *per se* is poorly represented in aquatic ones).

About half of all plant species, including food-producing crop species, are pollinated by animals. For example, the pollination of various fruit crops by bees and other insects is critical in mountain areas of Asia. In Nepal *Apis cerana* begins foraging at temperatures 5-7°C lower than those that initiate *Apis mellifera* foraging. Managed crop pollination with a variety of bee species in different zones plays an important role in overall agricultural development. The benefits of pollination are also considerable in high input-high output agriculture: the economic value of pollination services in the United States is estimated in billions of dollars per year. Management practices that reduce the species or abundance of pollinators can result in less genetic variation in crops dependent on pollinator visits for reproduction, both in temperate and tropical agriculture. With a loss in pollinators, seed production declines and the vulnerability to pests and climatic change increases with the resulting loss of genetic diversity.

"Mobile link" species (*i.e.* animals necessary for the persistence of plant species that in turn support otherwise separate food webs) such as pollinators and seed dispersal agents may be critical to the maintenance of the species richness of tropical forest based agroecosystems and complex home gardens imitating the natural forests' architecture. Many species in tropical forests managed for food and agriculture depend on a small suite of frugivores for dispersing their seeds. Loss of these species of fruit eaters may adversely affect the long-term viability of many tree species important for food security. Reductions in genetic variability are likely to be high for plant species that are highly dependent on frugivory for seed dispersal.

Biological diversity conservation functions. There is no strict divide between "wild" and "domesticated" species important for food and livelihoods. Many wild plant species and populations that have been considered to be wild are in fact carefully nurtured by people. A similar continuum exists for animal species that use agroecosystems as habitat, nesting grounds and food. Whilst not necessarily the subject of conscious management by herders or farmers, many wild species thrive in, or are dependent on, agroecosystems. In general the more structurally and biologically complex the agroecosystems, the more diverse the forms of wildlife. For example:

- The systems of shifting cultivation common in Asia have long been considered to be damaging to biological diversity in particular. However, given the opportunity, shifting cultivators preserve wild resources. Studies in Asia have concluded that most of the mature forests in this region are not virgin forests, but merely old forests that have reached a relatively stable dynamic state after some earlier clearing by human or natural means. Under traditional systems of shifting cultivation which speed up natural succession, wildlife flourishes, with elephants, wild cattle, deer, and wild pigs all feeding in the abandoned fields. Tigers, leopards and other predators are in turn attracted by the herbivores. Older fields contain a high proportion of fruit trees that attract primates, hornbills, squirrels and a variety of other animals.
- Long term research in the Lowland of west Poland shows that a mosaic of landscape structure of small cultivated fields, shelterbelts, meadows and small ponds helps maintain biological diversity (and many other functions such as enhanced water storage). Mammal,

amphibian, reptile, insect diversity are enhanced and the number of bird species is positively correlated with the diversity index of the landscape.

Climate functions. Although agricultural and the atmosphere interact in many ways, the links between climate and weather, and agricultural biodiversity can be rather complex.

At the global scale, the distribution of individual plants and animals, vegetation and crops is conditioned by available climatic resources such as solar radiation, which controls the production potential, and by rainfall, which determines to what extent the radiant energy can actually be used by plants for their growth. Indeed, it is for these reasons that climatic classifications largely coincide with vegetation maps. At the local scale, types of landscapes and vegetation, including crop and crop-vegetation mixes contribute towards modifying the local climates by directly affecting wind patterns, rates of evaporation, rainfall interception (effective rainfall), etc. This in turn conditions the development of vegetation and crop canopies which can be said to create their own microclimate, resulting from the interaction of plants with the general climate. Shelter belts, or the use or large tree belts to protect tropical plantation crops from cyclones, provide clear examples in which people and communities derive direct advantage from these landscape features.

Many examples demonstrate the benefits of maintaining minimum agrobiodiversity in the face of climate variability. Recent droughts in some southern African countries for example, have shown that mixes of local varieties planted over several weeks have the potential to better resist unusual patterns of rainfall variability (*e.g.* early-season drought) than some modern varieties planted on the "optimum dates". In fact, there appears to be a link between crop biodiversity and relatively low but regular production, one of the keys to food security.

Similarly, atmospheric pollution has been shown to interact with agro-ecosystem functions, both positively (nutrients, such as nitrates and sulphur) and negatively (toxic compound like tropospheric ozone and heavy metals). Given that the response to these substances varies enormously across the spectrum of species, they have the potential to modify patterns of biological diversity.

Agriculture also contributes towards the emission of some of the most significant "greenhouse" gases (in terms of global warming potential, GWP). Some studies have suggested for example, that increases in the area of permanently or quasi-permanently flooded rice cultivation account for a significant proportion of the increase in net methane emissions. The increased specialisation of ruminant production based on high yielding breeds has also significantly contributed to global methane emissions (methane is produced by anaerobic digestion in animals). An extreme case would be the enhanced biogenic emissions of carbonylsulphide associated with increasing cultivation of high sulphur crops such as rape. Oxidation of carbonylsulphide in the stratosphere leads to the production of sulphate aerosols that influence the intensity of ultra violet radiation reaching the earth's surface, potentially affecting the dynamic functions of agricultural biodiversity.

Volatile organic compounds such as terpenes and isoprenes are produced and released into the atmosphere by many plant species, especially in agroecosystems dominated by Mediterranean shrubs, eucalyptus and conifers. By influencing the oxidation capacity of the troposphere, these volatile compounds influence the abundance and distribution of trace gases such as ozone. In almost all cases however, many of the these effects are marginal when compared with the climatic effects of land use change, including deforestation, which have until recently constituted one of the main non-industrial sources of carbon dioxide

Functions in the water cycle. There are few experimental studies exploring the links between agricultural biodiversity and water. However, available evidence shows that agricultural biodiversity plays a crucial role in cycling water from the soil to the atmosphere and back. It also has measurable impacts on water quality. Agroecosystems with different species assemblages and plant architectures result in differences in the amount of precipitation intercepted, the proportion of precipitation converted to stem flow, and the proportion of precipitation that infiltrates the soil rather than running off. At the landscape level, the types, relative abundances and relative spatial locations of agro-ecosystem types affect the amount of water moving from one point to another. For example, conversion of vegetation within a watershed from forest or shrub land to less structurally and biologically complex grassland is known to influence stream flow out of the watershed, in both temperate and tropical systems.

At the functional group level, the root structure, phenology and physiology of different plant species important for food and agriculture have direct implications for the quantity and timing of water transfer to the atmosphere *via* evapotranspiration. Individual plant species differ in their resistance to water stress, their efficiency of water uptake from the soil, and so on. Genetic variations among crop varieties (differences in water use efficiency, stomatal resistance...) also influence these processes. At the landscape level, evapotranspiration from agroecosystems can have an effect on relative humidity and microclimate downwind.

In both terrestrial and aquatic environments, the plant and animal components of agricultural biodiversity also function to alter water quality by performing various filtration, uptake and excretory processes. These functions all affect the composition and concentration of dissolved gases, solutes and particulates. Species level differences in physiology can positively or negatively affect water quality. However, in most cases a greater diversity of biological organisms (from microbes to fish and macrophytes) leads to a higher quality of water for human consumption and use.

The influence of agricultural biodiversity on landscape structure. A landscape is a heterogeneous area made up of a cluster of interacting ecosystems that is repeated in similar form throughout. The spatial layout between landscape elements together with the interactions and linkages between them determine the landscape's structure and functions. The many different species found in a landscape are essential components of that landscape. By providing environmental services and functions agricultural biodiversity can have a profound influence on landscape structure. Through its positive or negative effects on agricultural biodiversity, human activity can transform whole landscapes over large areas.

For example, many rural communities enrich their agricultural plots and forest fallows with valued perennial plants. Through such enrichment practices, successional vegetation can become a site for economic production as well as for ecological rehabilitation. Each of the major tropical forest regions has many economic woody plants that have been managed, probably for millennia, in enriched fallows. In Vanuatu the natural composition of forests has been dramatically altered by centuries of itinerant gardening, favouring tree species that bear edible fruits and nuts. Fallows have been (and still are) enriched with rattan in East Asia, rubber in Sumatra, *Casuarina* in Papua new Guinea, *Gliricidia* and peach palm in Central America, oil palm in West Africa, and edible fruits and nuts universally. Locally adapted enrichments have altered species composition and also directly influenced the structure of landscapes at different spatial scales.

The influence of agricultural biodiversity on landscape structure is partly determined by the social institutions that mediate the relationships between rural people and the environment, and partly by climate and edaphic factors. For example in the semi-arid landscapes used by pastoralists of Africa, there are high levels of spatial and temporal variability in fodder biomass production, highly variable rainfall and episodic chance events such as drought. In these non-equilibrium systems pastoralists have developed opportunistic management schemes to exploit the patchiness of the vegetation, learnt to avoid risks by moving herds and flocks to make best use of heterogeneous landscapes and diversified their livelihood activities.

Pastoralists have rules and regulations which govern the use of water, pasture, animal movement and control of vegetation and trees. Management of agricultural biodiversity and the larger landscape is mediated by these local institutions. The local adaptive management of agricultural biodiversity enables people to cope with uncertainty and sustain the dynamic environmental processes that define and shape those landscapes. This is in stark contrast with the degradation that occurs under the centrally planned, standardised rangeland

and livestock management schemes often based on erroneous concepts of carrying capacity and equilibrium ecology. New perspectives in ecology have challenged the conventional views of drylands in Africa as stable ecosystems subject to decline and desertification once carrying capacity is exceeded. Rangelands and pastoral landscapes are resilient and less prone to degradation and desertification than once thought. The new findings concord with the knowledge of many local livestock herders and emphasise how rangelands are subject to high degrees of uncertainty and ecological dynamics, characterised by sudden transitions rather than slow and predictable change.

Specific components of agricultural biodiversity are often directly implicated in the processes that structure agroecosystems at different temporal and geographical scales (from small farm plots to whole water/ landscapes). Even highly complex landscapes like tropical irrigated rice or forests in the savannah transition zone of West Africa, are apparently structured by a very few key variables. Research over the past 20 years in applied ecology of managed systems shows that ecosystem and landscape dynamics tend to be organised around a small number of nested cycles, each driven by a few dominant variables.

A small number of plant, animal, and abiotic processes structure biomes over scales from days and centimetres to millennia and thousands of kilometres. Individual plant and biogeochemical processes dominate at fine, fast scales; animal and abiotic processes of mesoscale disturbance dominate at intermediate scales; and geomorphological ones dominate at coarse, slow scales....the physical architecture and the speed of variables are organised into distinct clusters, each of which is controlled by one small set of structuring processes. These processes organise behaviour as a nested hierarchy of cycles of slow production and growth alternating with fast disturbance and renewal.

Identifying and understanding the dynamics of these "structuring variables" provides a practical basis for sustainable agriculture and landscape management.

CELLULASE PRODUCTION

Successful utilization of cellulosic materials as renewable carbon sources is dependent on the development of economically feasible process technologies for cellulase production, and for the enzymatic hydrolysis of cellulosic materials to low molecular weight products such as hexoses and pentoses.

Spano et al. (4) showed that cellulase production, was the most expensive step during ethanol production from cellulosic biomass, in that it accounted for approximately 40% of the total cost. Significant cost reduction is required in order to enhance the commercial viability of cellulase production technology.

A cellulosic enzyme system consists of three major components: endo-ß-glucanase (EC 3.2.1.4), exo-ß-glucanase (EC 3.2.1.91) and ß-glucosidase (EC 3.2.1.21). The mode of action of each of these being:

- Endo-p-glucanase, 1,4-ß-D-glucan glucanohydrolase, CMCase, Cx: "random" scission of cellulose chains yielding glucose and cello-oligo saccharides.
- Exo-P-glucanase, 1,4-ß - D-glucan cellobiohydrolase, Avicelase, C1: exo-attack on the non-reducing end of cellulase with cellobiose as the primary structure.
- ß-glucosidase, cellobiase: hydrolysis of cellobiose to glucose.

Reese et al. 1950 (5) proposed that exo-ß-glucanase causes a disruption in cellulose hydrogen bonding, followed by hydrolysis of the accessible cellulose with endo-ß-gucanase. According to this hypothesis, in a synergistic sequence of events, endo-ß-glucanase acts randomly on the cellulose chain, while exo-ß-glucanase acts on exposed chain ends by splitting off cellobiose or glucose. Cellobiose is subsequently hydrolysed by p-glucosidase to glucose. This hypothesis is however the opposite of that proposed by Reese et al. (5), and indicates that three, rather that two enzymes are essential for the decomposition of cellulosic biomass.

SCREENING OF CELLULASE-PRODUCING MICROORGANISMS

Although a large number of microorganisms are capable of degrading cellulose, only a few of these microorganisms produce significant quantities of cell-free enzymes capable of completely hydrolysing crystalline cellulose *in vitro.* Fungi are the main cellulase-producing microorganisms, though a few bacteria and actinomycetes have also been recently reported to yield cellulase activity. Microorganisms of the genera *Trichoderma* and *Aspergillus* are thought to be cellulase producers, and crude enzymes produced by these microorganisms are commercially available for agricultural use.

Microorganisms of the genus *Trichoderma* produce relatively large quantities of endo-ß-glucanase and exo-ß-glucanase, but only low levels of ß-glucosidase, while those of the genus *Aspergillus* produce relatively large quantities of endo-ß-glucanase and ß-glucosidase with low levels of exo-ß-glucanase production.

In order to obtain a high titer cellulase-producing strain, various cellulase-producing strains obtained from Japanese and overseas research institutions were screened in our laboratory. On the basis of this screening, *Trichoderma reesei* QM-9414, isolated by the U.S. Army Natick Laboratory, was selected as the best cellulase-producing strain and was subjected to mono-colony isolation in order to obtain strain KY-746, which was used as a parent strain. A plate screening system which enables the isolation and evaluation of mutants was initially investigated, which led to the development of methodology for producing high titer cellulase.

Fungal strains of the genus *Trichoderma,* when cultured on an agar medium, are generally difficult to isolate, owing to the lateral spreading of hyphae over the surface. Hyphal growth was controlled by the addition of Triton X-100[7] and L-sorbose, an inhibitor of P-l,3-glucan biosynthesis, to the medium, thus enabling a large number of strains to be screened..

Large numbers of mutant strains exhibiting a diversity of characteristics were obtained by UV and nitrosoguanidine (NTG) treatment and various other techniques. A strain free from catabolite repression for example, was induced using a culture medium containing glycerol and glucose, while a high titer strain was induced by addition of L-sorbose to the medium. In addition, a strain of elevated (3-glucosidase activity was isolated from medium containing 4-methyl-umbelliferyl-ß-D-glucoside.

STRAIN IMPROVEMENT FOR CELLULASE PRODUCTION

The commercial use of cellulases is dependent on the following:

- High titer and good enzymatic activity
- Low production cost, and
- Feasible mass production.

DEVELOPMENT OF A PROCESS FOR HIGH-TITER CELLULASE PRODUCTION

Representative mutants derived from KY-746 were batch cultured in a 5-L fermentor. Considerable enhancement of titer was observed in four strains: KDG-3, PC-3-7, PCD-10, and CDU-11. An evaluation of carbon source utilization by these microorganisms, suggested that the production of high-titer cellulases necessitated an increase in the concentration of Avicel, the carbon source.

A somewhat lower than expected maximum cellulase titer was however obtained at Avicel concentrations greater than 6% in batch culture. This relatively lower titer, may be attributed either to adsorption of the cellulase produced on to the Avicel, or rate limitation of aeration due to the high viscosity of the culture medium.

In order to overcome this, semi-batch culturing was investigated, where it was found that an initial Avicel concentration of 6%, followed by subsequent addition of 4% Avicel together with other components after a few days, resulted in relatively higher titer cellulase than that obtained in the batch process. Titer improvement by the semi-batch process was however dependent on the types of microorganisms used.

While the titer of KY-746 was relatively unaffected by the semi-batch process, various degrees of improvement in titer were observed for KDG-3,

PC-3-7, PCD-10, and CDU-11. This semi-batch process was further utilized in order to optimize the stirring rate, aeration volume, feed change, seed volume, and the seed culture period.

T. reesei generally exhibits poor (3-glucosidase activity. Strain CDU-11 was therefore created in order to enhance p-glucosidase activity. P-glucosidase activity can also be induced by elaborately controlling culture conditions.

CMCase, Avicelase, and FPU activities exhibited a pH optimum of approximately 4, while the pH optimum of P-glucosidase was between pH 5 and 6. pH adjustment was therefore investigated with the objective of uniformly inducing all three activities.

Culturing at pH 4 for 2 days, followed by a shift to either pH 5 or 6 on the third day, resulted in a marked increase in the p-glucosidase activity of PC-3-7, but no beneficial effect was observed in the case of CDU-11 and of KDG-3, which was already of high [3-glucosidase activity. On the basis of these findings, a scale-up experiment was conducted in a 1-kL tank using the semi-batch culturing method with 10% Avicel as a carbon source and soybean curd as a nitrogen source.

The pH was controlled and shifted as described above.. With a shortened culture period of 12 days, elevated activities of 530 U/ml and 8 U/ml were obtained for CMCase and p-glucosidase respectively, with an elevated protein content of 40 mg/ml in the culture broth. The soluble protein content of the culture broth was 53 mg/ml, which was considered favourable.

PRODUCTION OF AMINO ACIDS

Amino acids play central roles both as building blocks of proteins and as intermediates in metabolism. The 20 amino acids that are found within proteins convey a vast array of chemical versatility. The precise amino acid content, and the sequence of those amino acids, of a specific protein, is determined by the sequence of the bases in the gene that encodes that protein. The chemical properties of the amino acids of proteins determine the biological activity of the protein. Proteins not only catalyse all (or most) of the reactions in living cells, they control virtually all cellular process. In addition, proteins contain within their amino acid sequences the necessary information to determine how that protein will fold into a three dimensional structure, and the stability of the resulting structure.

The field of protein folding and stability has been a critically important area of research for years, and remains today one of the great unsolved mysteries. It is, however, being actively investigated, and progress is being made every day. As we learn about amino acids, it is important to keep in mind that one of the more important reasons to understand amino acid structure and properties is to be able to understand protein structure and properties. We will see that the vastly complex characteristics of even a small, relatively simple, protein are a composite of the properties of the amino acids which comprise the protein.

ESSENTIAL AMINO ACIDS

Humans can produce 10 of the 20 amino acids. The others must be supplied in the food. Failure to obtain enough of even 1 of the 10 essential amino acids, those that we cannot make, results in degradation of the body's proteins—muscle and so forth—to obtain the one amino acid that is needed. Unlike fat and starch, the human body does not store excess amino acids for later use—the amino acids must be in the food every day.

The 10 amino acids that we can produce are alanine, asparagine, aspartic acid, cysteine, glutamic acid, glutamine, glycine, proline, serine and tyrosine. Tyrosine is produced from phenylalanine, so if the diet is deficient in phenylalanine, tyrosine will be required as well.

The essential amino acids are arginine (required for the young, but not for adults), histidine, isoleucine, leucine, lysine, methionine, phenylalanine, threonine, tryptophan, and valine. These amino acids are required in the diet. Plants, of course, must be able to make all the amino acids. Humans, on the other hand, do not have all the enzymes required for the biosynthesis of all of the amino acids.

WHY LEARN THESE STRUCTURES AND PROPERTIES?

It is critical that all students of the life sciences know well the structure and chemistry of the amino acids and other building blocks of biological molecules. Otherwise, it is impossible to think or talk sensibly about proteins and enzymes, or the nucleic acids.

AMINO ACID PRODUCTION

The amino acid business is a multi-billion dollar enterprise. All twenty amino acids are sold, albeit each in greatly different quantities. Amino acids are used as animal feed additives (lysine, methionine, threonine), flavour enhancers (monosodium glutamic, serine, aspartic acid) and as speciality nutrients in the medical field.

Glutamic acid, lysine and methionine account for the majority, by weight, of amino acids sold. Glutamic acid and lysine are made by fermentation; methionine is made by chemical synthesis. The major producers of amino acids are based in Japan, the US, South Korea, China and Europe. Many microbe-based industries have their origins in traditions that go back hundreds or thousands of years. The amino acid industry has its roots in food preparation practices in Japan. Seaweeds had been used for centuries there and in other

Asian countries as a flavouring ingredient. In 1908, Kikunae Ikeda of Tokyo Imperial University isolated the flavour enhancing principle from the seaweed *konbu* (also spelled *kombu, Laminaria japonica;* related to kelp) as crystals of monosodium glutamate (MSG).

Adding MSG to meat, vegetables and just about any other type of prepared food makes it savory, a property referred to as *umami*. Soon after Ikeda's discovery, and recognizing the market potential of MSG, Ajinomoto Co. in Japan began extracting MSG from acid-hydrolyzed wheat gluten or defatted soybean and selling it as a flavour enhancer. The production of MSG via "fermentation" grew out of the ashes of WWII in Japan. Around 1957, Japanese researchers led by S. Kinoshita at Kyowa Hakko Kogyo Co. isolated soil bacteria that produced large amounts of glutamic acid.

Producing strains were found by inoculating soil isolates in a grid pattern on duplicate Petri plates. The colonies were allowed to grow and one set of duplicates was killed with UV irradiation. The killed plate was overlaid with soft agar containing a *Leuconostoc mesenteroides*. Since *L. mesenteroides* required glutamic acid for growth, it only grew in the vicinity of colonies that had excreted glutamate. Potential glutamate producers were then picked from the duplicate, unkilled, plate.

Members of the *Actinobacteria* in the genus *Corynebacterium* (originally named *Micrococcus glutamicus*) were the most effective producers. Over the years, various glutamate-producing bacteria have been isolated and classified as *Arthrobacter, Brevibacterium,* or as members of other genera, but recent work has shown that almost all of these strains belong to the genus *Corynebacterium.* Wild-type cultures produced up to 10 g/l glutamic acid. Yields were quickly improved by process engineering and by developing over-producing mutants. Yields are now in excess of 100 g per liter.

The isolation of bacterial glutamate-producers led to the development of large-scale manufacture of MSG from cheap sugar and ammonia rather than from more expensive extracts of plants or animals. In the early 1960s, workers at the same company found that *C. glutamicum* homoserine auxotrophs produced lysine thus providing the first viable fermentation process for lysine production. Today these bacteria produce well over 1,000,000 metric tons of MSG and 600,000 metric tons of lysine annually.

Three general approaches are used today for making amino acids: direct chemical synthesis, fermentation and bioconversion using enzymes. Choosing between processes depends on available technology, costs of raw material, market prices and sizes, cost of running fermentation versus synthesis reactions, and the environmental impact of the process itself. The processes have not changed very much since then, but the amounts made have increased at a rate of about 2-5% per year.

Amino acids are consumed in a variety of markets. The largest by volume is the food flavouring industry. MSG, alanine, aspartate, arginine are all used to improve the flavour of food. The basis of our ability to enjoy the taste of amino acids is rooted in evolution. Animals require certain amino acids for growth and nutrition. Specific taste receptors on the human tongue have been found that are G-protein coupled receptors responsive to many of the 20 L- but not D-amino acids. For human receptors, the strongest response is for glutamate, thus providing a biochemical link to what cooks have known for centuries. Interestingly, inosine monophosphate (IMP), another well-known flavour enhancing nucleotide also sold commercially, greatly increases the response of the receptor to amino acids.

The second largest consumer of amino acids is the animal feed industry.

Lysine, methionine, threonine, tryptophan and others improve the nutritional quality of animal feeds by supplying essential amino acids that may be in low abundance in grain. Using 0.5% lysine in animal feed improves the quality of the feed as much as adding 20% soy meal. In addition, by limiting the added amino acid supplements to those required by the animal, some of the excess ammonia made via deamination reactions that is normally excreted to the environment is eliminated. The addition of microbially-produced amino acids has also increased with the onset of "Mad Cow" syndrome (BSE), a disease traced to feed supplements containing animal protein.

Finally, the pharmaceutical industry uses a variety of amino acids for making intravenous nutrient solutions for pre- and post-operative care. These mixtures account for a small percentage of the total volume of amino acids sold each year, but the requirement that they be highly purified provides a value-added component. In addition to these uses, amino acids are used as speciality chemicals in laboratories, in the manufacture of artificial sweeteners (aspartame) and in many other situations.

MSG Glutamate is the most abundant free amino acid in bacterial cytoplasm. Nevertheless, in order to be useful, glutamate producers must do two things well: they must overproduce glutamate in excess of their normal metabolic needs, and they must excrete it into culture broth. The precise mechanism by which *C. glutamicum* does these things is still not completely understood despite over forty years of study. Some physiological traits, however, are clearly involved. These include biotin auxotrophy of producing strains, a marked decrease in α- ketoglutarate dehydrogenase activity during production, and a predilection for exporting glutamate, perhaps via a specific transporter.

Many of the original glutamate-excreting strains were biotin auxotrophs, and growing in biotin deficient medium was found to "trigger" glutamate production. Biotin is a cofactor (a "vitamin") used by enzymes that carboxylate substrates. One such enzyme is acetyl-CoA carboxylase that converts Acetyl-CoA + CO_2 to Malonyl-CoA in the first step of fatty acid biosynthesis. Biotin auxotrophs growing in biotin deficient medium were proposed to have altered membranes due to suboptimal fatty acid biosynthesis. Supporting the notion of altered permeability is the observation that growth at higher temperatures, or including detergents or cell-wall biosynthesis inhibitors like penicillin in the growth medium can also trigger excretion.

Reduced levels of α-ketoglurate dehydrogenase during production may also be linked to membrane integrity. In corynebacteria, the enzyme has three activities on two peptides: the α-kg dehydrogenase + dihydrolipoamide S-succinyltransferase peptide and the dihydrolipoamide dehydrogenase peptide. The latter is shared with pyruvate dehydrogenase and is likely to be membrane bound and thus prone to being affected by trigger factors that alter membrane composition.

Since α-ketoglurate dehydrogenase catalyses a step in the TCA cycle, the cycle is largely incomplete during glutamate production, a circumstance requiring that pools of oxaloacetate be filled, as carbon is lost through glutamate. Anaplerotic enzymes replace OAA; these enzymes include pyruvate carboxylase (Rxn. 1), malic enzyme (Rxn. 2), PEP carboxylase (Rxn. 3) and glyoxylate pathway enzymes.

- Pyruvate + CO_2 + ATP → oxaloacetate + ADP + Pi
- Pyruvate + CO_2 + NADPH → malate + NADP+ + H+
- Phosphoenol pyruvate + CO_2 → oxaloacetate + Pi.

The actual contribution of each anaplerotic enzyme or pathway depends on the growth conditions, production phase and a host of interacting metabolic signals. However, mutants lacking pyruvate carboxylase produce as much glutamate as wild type suggesting that PEP carboxylase is the major anaplerotic enzyme whan growing on sugars. When growing on acetate or fatty acids, the glyoxylate pathway assumes a major role in filling OAA pools, although acetate is not normally used in the industrial process.

For many years, increased membrane permeability was thought to promote glutamate excretion. Recently, however, export via a specific efflux transporter has been proposed as a mechanism for avoiding exceptionally high levels of intracellular glutamate. Such an exporter has been proposed by analogy with the LysE transporter that is responsible for lysine and arginine export in *C. glutamicum.*

Other factors contributing to glutamate overproduction include metabolic flux alterations based on cell growth limitations. The following chain of events seems likely: A triggering mechanism (biotin depletion, high temperature, cell membrane alterations by detergents, oleic acid or antibiotics) results in a decrease or repression of α-kg dehydrogenase. The result is a redistribution of metabolites at the branch point in the TCA cycle leading from α-kg to succinyl-CoA or glutamate. The increase in glutamate levels, beyond that needed for cell growth, stimulates glutamate efflux as the cell attempts to maintain the proper level of intracellular glutamate. The result is the excretion of glutamate from the cell.

Industrial production of MSG Although the details of MSG production remain proprietary and differ in detail from company to company, a general outline of the industrial process is known. The process is most commonly run as a fed-batch type, where sugar is added during the fermentation process. The reason for using fed-batch rather than batch, where all components are present at the beginning of the process, is that high sugar concentrations, more that 20% in total would otherwise need to be added to the medium. Such high levels can lead to incomplete oxidation of the sugar to lactic or acetic acids or to osmotic inhibition of growth with subsequent decrease in yields.

Small seed cultures are grown on glucose, potassium phosphate, magnesium sulfate, yeast extract and urea as a source of nitrogen. Larger cultures use cheaper sources of sugar including cane or beet molasses and starch hydrolyzates from corn or cassava. The sugar source roughly parallels the geographic location of the process.

That is, corn syrup is used in the United States, tapioca (from cassava) in South Asia and cane and beet molasses in Europe and S. America. Ammonia and ammonium sulfate are used as nitrogen sources. Ammonia can also be used to control pH during fermentation. Cheap sources of vitamins and other nutrients include corn steep liquor, a by-product of cornstarch manufacture that is replete with amino acids, nucleic acids, minerals and vitamins.

To begin the process, stocks of *C. glutamicum* are used to inoculate shake flask cultures. The resulting cells are transferred to a small seed culture tank that is grown and used to inoculate a larger tank, and so on. The intermediate seed culture volumes are variable but generally in the range of 200-1000 liter, then 10,000-20,000 l and finally the production tank of about 50,000-500,000 l. The process is carefully controlled at each step such that cell density, nutrient composition, temperature, pH, aeration, agitation rates and sugar flow rates are as consistent as possible from batch to batch. Oleic acid (0.65 ml/l) might be added at the beginning of the fermentation to encourage glutamate excretion; pH is set around 8.5 with ammonia, and maintained at 7.8 during the process.

After 14 hours of growth, the temperature is increased from about 32-33°C to 35° Sugar is fed in as the fermentation proceeds up to about 36 h. During the course of the fed-batch process about 160 g or more of glucose per liter, or its equivalent, is fed into the bioreactor. Glutamate concentrations are monitored at intervals and, after the process is complete, the broth is pumped from the bioreactor to recovery tanks. Yields of glutamate from a large-scale fermentation are in excess of 100 g/l. That means that a 100,000 liter bioreactor yields about 10,000 kg of glutamate. At a market price of about $1.25 per kg, a single fermenter has a value of about $12,500. Considering the costs of energy, personnel, overhead and processing, that is not much money and is a consequence of fierce competition between companies.

L-LYSINE

Soon after *C. glutamicum* was commercialized for MSG production, auxotrophs were developed that excreted commercial quantities of lysine. By 1958, Kinoshita and colleagues had shown that lysine was excreted by homoserine auxotrophs. For example, a homoserine auxotroph (strain ATCC13287) was patented in 1961 that yielded 44 g lysine per liter with a conversion efficiency of 26% from sugar (g lysine/g sugar). Further increases in yield were made incrementally with the introduction of additional amino acid and vitamin auxotrophies plus development of strains resistant to antimetabolites. Typical strains today provide conversion efficiencies of over 50%.

To understand why lysine is overproduced from a physiological standpoint, one must first understand the regulatory circuits that govern lysine biosynthesis. The approaches used for creating amino acid overproducers are paradigms of rational strain improvement as contrasted with approaches that rely on random mutagenesis and screening. The pathway leading to lysine (also threonine, isoleucine, methione) biosynthesis is initiated with the conversion of aspartate to aspartyl-P via the enzyme aspartokinase (AK). The phosphorylated aspartate is then converted to aspartyl-semialdehyde (ASA) that can converted to homoserine by homoserine dehydrogenase (HSD) or to diaminopimelic acid (DAP) by a series of five enzymatic conversions, and thence to lysine. Homoserine is the starting point for making threonine and isoleucine as well as methionine. The major control points for the metabolic flux to individual amino acids occurs at the level of aspartokinase and homoserine dehydrogenase.

AK is feedback inhibited by the accumulation of threonine plus lysine. Homoserine dehydrogenase is feedback inhibited by threonine and its synthesis is repressed by methionine. Mutations that lead to the overproduction of lysine include HSD-, HSDleaky, and lysine analogue resistant mutants. HSD-mutants are homoserine auxotrophs that effectively lack the ability to make threonine, methionine and isoleucine, thus eliminating the feedback inhibition of AK by threonine. HSDleaky mutants make an HSD that is less effective at making homoserine so that threonine does not accumulate to levels sufficient to shut down AK. The benefit of the latter mutants is that additional amino acids do not need to be added to growth media to make up for their complete lack of synthesis in HSD null mutants. Analogue resistant mutants resist the feedback inhibition effects of allosteric effectors.

In the case of AK, mutants were derived that resisted analogues of lysine and/or threonine. To derive such mutants, mutagenized cultures are incubated in the presence of the analogue; those that grow "resist" the analogue and are screened for their ability to make lysine. The rationale behind the approach is that wild-type AK recognizes the lysine analog 2-aminoethyl-L-cysteine (AEC) as lysine. The enzyme is inhibited as if lysine had accumulated in the cell and the cell fails to grow due to lysine starvation. Analogue resistant mutants have AKs that do not recognize AEC as lysine; these enzymes remain uninhibited, continue to make aspartyl-P and then lysine and the cells continue to grow. Such mutants, since they no longer respond to intracellular lysine levels, continue to synthesize lysine at high levels. Other types of auxotrophs, for example for leucine and alanine also overproduce lysine since their biosynthetic pathways compete for pyruvate.

In 1996, the gene encoding a membrane protein involved in the active export of lysine– *lysE* – was cloned and sequenced. It is responsible for exporting lysine up a concentration gradient eventually leading to the accumulation of as much as 1 mole per liter of lysine. Mutations in this gene cause lysine to accumulate in the cytoplasm but not in the extracellular medium. Introducing the gene into a multi-copy plasmid in *C. glutamicum* leads to increases in the rate of efflux of intracellular lysine.

Lysine Fermentation The industrial production of L-lysine is carried out in a manner similar to that for L-glutamate production.

Sugar cane or beet molasses is used as the substrate for production. If the biotin level is low in the starting material it must be added to a level of about 30 ug/l. If an auxotroph for any amino acid is used, that amino acid must be present so that the cells may grow but should not be present at high enough concentration to exert inhibitory physiological effects such as feedback inhibition of the pathways used for production.

PRODUCTION OF ALGAL BIOMASS

Algae (cyanobacteria and unicellular eukaryotes) grow autotrophically and synthesize their food by taking energy from sunlight or artificial light, carbon source from carbon dioxide, and nutrients from carbohydrates present in growth medium. In a few countries, cultivation of algae is carried-out in large trenches *i.e.* particularly in sewage oxidation ponds by using sunlight or in an artificial illumination conditions for use in life supportive systems for extended space exploration.

Chlorella strains are being used for a variety of applications in biotechnology. Due to their very high protein contents, they serve to improve protein deficiency and can be used as feed for production of animal protein. In many countries strains of *Chlorella* are utilized for sewage oxidation and waste water treatment.

For cultivation of algae on sewage wastes, oxidation ponds are prepared, where sewage is allowed to accumulate. It is awaited till mixed cultures of algae grow (or inoculated with a singly prepared algal culture). For example, in Japan mixed culture of *Chlorella ellipsoides* and *Scenedesmus obliguus* was developed in open pond systems.

FACTORS AFFECTING BIOMASS PRODUCTION

Following are the factors affecting the yield of biomass:

- Illumination time;
- Light intensity;
- Supply of CO_2. Concentrations of CO_2 differ in different conditions, for example, an alkaline lake. Lake Texcoco in Mexico, has high concentration of sodium carbonate. On the other hand, algal growth is limited as a result of liberation of CO_2 and ammonia by bacterial activity;
- Nitrogen sources (ammonium salts or nitrates are the suitable nitrogen sources which increase biomass yield);
- Agitation of growing cells to maintain cells in suspension.

Biomass yield ranges between 12 and 15 $g/m^2/day$ (on dry weight basis) as obtained with *Spirulina maxima* and *Scenedesmus quadricauda* grown in out door pond conditions.

Mass cultivation of algae has been started in many countries, such as Japan, West Germany (now Germany), Mexico, Chechoslovakia, India, etc. In India, National Environmental Engineering Research Institute (NEERI), Nagpur has developed a technique of cultivating algae in sewage oxidation pond systems.

This practice is also in use at NBRI (Lucknow), Hyderabad and other centres. Interestingly, experiments conducted at the CFTRI, Mysore have shown that the microalgae *e.g. Scenedesmus acutus* and *Spirulina platensis* could be cultivated on a large scale and used as food and feed as they are rich in protein and their nutrient value is comparable to conventional foods.

HARVESTING THE ALGAL BIOMASS

Harvesting of algal culture becomes problematic because of settling down of cells at bottom and mixing of the algal cultures. The cells are recovered by concentration, dewatering and drying. Sometimes flocculants *e.g.* aluminium sulfate and calcium hydroxide and cationic polymers are added to the medium but they cannot be separated from the harvested cells. Therefore, this method warrants the application of SCP products in food and feed.

Methods of separation and concentration also follow centrifugation, flocculation and centrifugation plus flocculation, but it is not economically feasible so far.

Harvesting the cyanobacteria, for example, *Spirulina* sp. is less troublesome as their spiral filaments float on the surface of water because of gas filled vacuoles in their cells which result in floating algal mats. Cells are able to fix atmospheric nitrogen. Algal mats are filtered and suspension of *Spirulina* is dried with hot air to get fine powder. Algal yield from stabilization pond is around 114 tonnes/ha/year. In California 70 tonnes/ha/year of *Scenedesmus* was obtained from sewage. From Bankok much high yield of about 170 tonnes/ha/year has been reported.

SCP PRODUCT OF SPIRULINA

In 1821, Bernal Diaz del Castillo, for described the biscuits with the name 'tecuitlatl'. It contained dried mats of *S. maxima* collected from the Lake Texcoco. The biscuit was sold in the markets of Mexico. In 1964, J. Leonard (a Belgian botanist) took part in an expedition and noticed dried biscuits of blue green colour in several village markets. The biscuit *dike* consumed by local population consisted of *Spirulina platensis.* Many pilot plants for the production of *Spirulina* powder have been established in Japan, United States and European countries. Sosa Texcoco is the first Mexican Company to set up the first pilot plant in 1973 which produced about 150 tonnes *Spirulina* powder per annum; but the yield was increased to 1,000 tonnes in 1982. This company exported powder to the United States and prepared lozenges and capsules from the powders by adding vitamin A and C. The Mexican Company also supplied *Spirulina* powder to government institutions which were in charge of improving the nutritional situation of the population. Institutions used *Spirulina* powder to make biscuits and confectionery with a high protein content.

BENEFITS FROM *SPIRULINA* SCP

Mass cultivation of *Spirulatina* offers several advantages over *Chlorella* and *Scenedemus* as given below:

- Being a filamentous alga, *Spirulina* can be harvested by simple and less expensive methods such as nylon or cotton cloth filter.
- Filaments of *Spirulina* float on water surface due to presence of gas vacuoles. Hence, there is no problem of harvesting unlike *Chlorella* and *Scenedesmus.*
- There is least chance of contamination in growth tanks of *Spirulina* as it grows at high alkaline pH 8-11.
- Heat drying is sufficient for *Spirulina* as it has thin cell wall, whereas

spray drying is required for *Chlorella* and *Scenedesmus* which is expensive,

- Researches done by UNIDO programme in Mexico (1980), Mexican National Institute of Nutrition, Hyderabad (1988, 1990) under Indian Council of Medical Research on several aspects of possible adverse changes in multigeneration feeding tests on laboratory animals and humans have shown no adverse effects.
- *Spirulina* is highly digestive (85-95%) due to thin wall and low nucleic acid contents (4%). It contains high percentage of digestible proteins (62-72%), vitamins, amino acids and other nutrients. The aminogram of *Spirulina* is comparable to the FAD, milk and egg protein pattern.

STATUS OF FOOD DISTRIBUTION IN INDIA

Food accounts for the largest share of consumer spending. Food and food products account for about 53% of the value of final private consumption estimated at INR 8600 bn (2003-04 at current prices). Food products are sold through over 5 million food and grocery stores in India. The size of the organised food retail segment in India is estimated at INR 20 bn (approximately 0.2 % of food expenditure).This is in contrast to most of the developed countries, where food distribution is highly consolidated.

In countries such as Australia, France, Germany, Denmark and The Netherlands, the share of top the 10 retailers is more than 80% (USA ~ 35%, China ~ 15%).

Most organised retailers in India are regional and use single formats such as convenience stores, supermarkets, hypermarkets or cash and carry. In contrast, most international retailers have multiple format models. Besides reflecting the stage of evolution of retailing in India, this also highlights the capital constraints of existing players in India, restricting their ability to make large investments.

In India, food companies are much larger in size than the organised food retailers in India. In contrast, the retailers are larger in scale, than food companies, in developed countries. Given the channel power of retailers in these markets, food companies have started establishing partnerships with them to develop and test new products, share consumer information and undertake consumer promotions.

SOURCES OF FINANCE FOR AGRICULTURE & FOOD PROCESSING

At present nearly 60% of rural credit is used for agriculture and allied activities, 10% for non-farm activities and the balance 30% for funding household consumption expenditure. Nearly 40-45% of rural credit needs are catered to by formal credit institutions and the balance by the informal sector including commission agents, input suppliers, traders, Self Help Groups (Joint Liability Groups), processing industries and professional money lenders. The Indian commercial banks are mandated by the Reserve Bank of India (RBI), to undertake directed lending for agriculture and rural development. Commercial banks are required to achieve priority sector lending of 40% of net bank credit of which 18% should be for agriculture. Further, sub-targets are also specified *i.e.*, Direct Agriculture (credit for direct farmer benefit, Target: Minimum of 13.5% of the net bank credit) and Indirect Agriculture (Target: Maximum of 4.5% of the net bank credit).

In FY 03, total agricultural advances outstanding under priority sector lending, by Public Sector Banks stood at INR 735 billion, on a total net bank credit portfolio of INR 4779 billion. As compared to this, agricultural advances by Private Sector Banks stood at INR 119 billion, on a total net bank credit portfolio of INR 718 billion. The following Exhibit provides details of direct and indirect priority sector credit amongst Public and Private sector banks during FY03.

In case of a shortfall in lending to agriculture, banks are required to invest in Rural Infrastructure Development Fund (RIDF) deposits, which offers interest rates linked to the bank's performance in lending to agriculture. These rates are inversely proportional to the shortfall in agricultural lending. The interest rate on RIDF is provided in slabs, based on the achievement of the priority sector lending target, and can be as low as 4% to 5%.Therefore, it is imperative to achieve the agriculture target and more specifically the Direct Agriculture target as it accounts for 75% of the total.

Established in 1996, RIDF is being utilized for making investments in rural infrastructure projects. Nine tranches of RIDF have been established with an aggregate corpus of INR 340 billion, with the ninth tranche of INR 55 billion. Cumulative sanctions and disbursements under various tranches of RIDF stood at INR 334 billion and INR 193 billion respectively, at the end of February 2004.The RIDF funds are used to finance minor irrigation projects, shallow wells, drinking water facilities, roads, bridges and storage infrastructure.

SOURCE OF FINANCE-FOOD PROCESSING

Banks: The Food Processing sector has access to credit from Commercial Banks-Indian and Foreign, Cooperative Banks and the Regional Rural Banks. The formal institutional sector assists the food processing companies through long term loans for capital investments and short term loans for working capital.

The total lending to the food processing sector is clubbed with lending to the agriculture sector and separate details for financing to the food processing sector are not available.

- National Bank for Agricultural and Rural Development (NABARD): NABARD offers refinance facilities for food processing, agri infrastructure, developmental assistance for RRBs, DCCBs, SSI and SHG linkages, and assists in infrastructure development and research.
- Small Industries Development Bank of India (SIDBI):SIDBI has been involved in assisting the entire SSI Sector including tiny, village and cottage industries through suitable schemes tailored to address the funding requirements for expansion, diversification, modernisation and rehabilitation.
- Export Import Bank (EXIM Bank): Exim Bank assists in financing and facilitation of foreign trade. The Bank also offers financial support to companies engaged in exports.
- National Cooperative Development Corporation (NCDC): NCDC assists in promotion, planning and financing the agricultural supply chain from production, processing, storage and trade of agricultural produce and food products. NCDC also provides assistance for marketing of certain notified commodities *e.g.* fertilizers, pesticides, agricultural machinery etc.
- Ministries/Government bodies
- Ministry of Food Processing Industries (MFPI): MFPI is the Nodal agency for development of the processed food sector in the country MFPI's financial schemes include schemes for technology upgradation, Human resource development, Quality testing, R&D,TQM, backward and forward integration, development of infrastructure including food parks.
- Agricultural and Processed Foods Products Export Development Authority (APEDA): APEDA facilitates market linkages between Indian producers, manufacturers and the international market. APEDA provides financial assistance for market development, infrastructure development and development of quality enhancing facilities.
- Ministry of Agriculture (MoA), Government of India: The Ministry of Agriculture under various schemes, provides financial assistance for development of specific crops for investment in seeds, irrigation, farm implements, inputs, infrastructure and training.
- National Horticultural Board (NHB):NHB promotes integrated development in horticulture, assists in development of post harvest management infrastructure, promotes production and processing of fruits and vegetables, strengthening of market information systems

and assists in R&D programmes in cultivation and processing. NHB's financial schemes are directed towards commercial horticulture and infrastructure related to post harvest techniques.

Financial assistance from these organizations are in the form of grants, back-ended subsidies, soft loans, refinance etc., with most of the schemes directed to specific sub-sectors of the agri/food processing industry.

In addition, some State Governments offer financial schemes to companies in the sector. The key issues with Government schemes are highlighted below:

- Overlap among schemes: Various Government agencies have separate schemes which aim to achieve the same objectives. Thus, an applicant can potentially approach different agencies and obtain assistance from more than one agency. Some of the examples of overlapping schemes are as follows:
- Several areas addressed inadequately: There are a host of areas which are not addressed by these schemes such as financing of freezer cabinets for retail outlets, financing of bulkcoolers for milk, financing of alternate markets/mandis, financing of vending machines for tea/coffee/beverages etc.
- Creation of excess capacity: There are instances where assistance has been provided to set up cold storages in locations which have excess storage capacity. This has lead to an unviable situation for all cold storage operators in the region. It is essential for the nodal agencies to achieve development of the food processing sector in a sustainable manner.
- Inadequate financial assistance^ most instances, there is a cap on total financial assistance provided at INR 5-5.5 mn. This leads to fragmentation in capacity creation, and often, a situation of overcapacity. At the same time, it does not allow players to scale up their operations.
- Inadequate Monitoring of progress of schemes: It is important to track progress of the projects funded by the Government. Periodic appraisal of the progress will ensure that funds are appropriately utilized. This will also enable assessment of success and lacunae of these schemes and record the multiplier effect of the financial support provided.
- Most of the schemes are back-ended which leads to a funds crunch during implementation.
- Institutions providing subsidies do not undertake their own assessment and rely on the appraisal report of the bank which provides loans to the project.

- There are significant time lags from the date of application for financial assistance, to release of funds, and may affect the project schedule, and lead to cost overruns
- These schemes do not address working capital requirements, which due to the nature of the industry are significant and often higher in terms of quantum than capital investment requirements
- Applicants claim that they are often unaware that their applications have been rejected on account of lack of communication from the funding institution. Further, applicants are unaware of the reasons for rejection.

FINANCING NEEDS OF THE AGRICULTURAL AND FOOD SUPPLY CHAIN

The Indian agribusiness supply chain is highly fragmented, with independent players engaged in various value-addition activities such as input supplies, cultivation, processing, distribution, storage and retailing. The scale of operations of each entity is limited not only due to the structure of the supply chain, but also on account of Government policies as well as due to demand-related factors.

In contrast, internationally, a large number of food companies are integrated across the chain. *e.g.* players such as ADM and Cargill have business interests in agricultural inputs, procurement, transportation and storage of produce, processing and value addition. The structural complexion of the Indian supply chain thus translates into limited scale of financing as well as higher risk, given the lack of control of each of the players, on the supply chain.

ISSUES WITH FARMER FINANCING

It is important to understand the issues faced in farmer level financing, both from the borrower's as well as the lender's perspective, and develop solutions which address these issues comprehensively.

Non-availability of timely credit from organized sources, mainly on account of the processes and procedures of banks/financial institutions. This compels farmers to rely on intermediaries.

- Inadequacy in credit availability-The 'scale of finance' method, stipulated by RBI, caps total credit availability per farmer.
- Inability to offer tangible security/collateral cover to access credit from banks Subsistence farmers, who constitute a large proportion of India's farming community, are unable to benefit from the organized credit system since they cannot offer adequate tangible security. Further, the security offered (land documents) are notional and cannot be enforced by banks.

- The above factors, thus translate into continued reliance on intermediaries for financing. While intermediary financing is timelyas they do not require the level of documentation as banks, the cost of funds is exorbitant, nearly 4 times that of interest rates charged by banks.
- A more efficient agri-supply chain requires that farmers' financing needs are addressed by banks and linkages of farmers with middlemen are scrapped.

Banker's Perspective

- Lack of credit discipline among borrowers Lack of credit discipline, to some extent is infused by the Government's decision to waive outstanding loans of farmers. Policy-related uncertainties translate into a huge risk for banks. Instead of loan waivers, the Government should seek to impose a short-term moratorium on repayments, to enable farmers to overcome the adverse impact of a poor monsoon/ crop failure.
- Lack of tangible security from farmers-as mentioned above, farmers are unable to provide tangible security to financiers.
- Lack of linkages of farmers with processors-In absence of forward linkages with markets, banks run a huge performance risk, as there is greater potential for farmer default.
- Inadequate insurance cover the existing crop insurance schemes are inadequate, and this further increases the risk profile of farmers
- Norms for priority sector lending-As per the current RBI policy on priority sector lending, kharif credit is not accounted for in priority sector lending. This results in a lop-sided focus on extending financing to farmers, favouring crops cultivated in the rabi season.

Policy-related Issues

- Cooperatives Act: There are about 95,000 Primary Agriculture Credit Co-operatives (PACS) and about 100,000 marketing co-operatives. As per the Cooperative Act, cooperatives are allowed to borrow from Cooperative Banks, DCCBs and RRBs. This provision needs to be extended to scheduled banks, in order to enhance flow of credit to farmers.
- State Warehousing Corporations Act: Private sector scheduled banks are constrained from lending to State Warehousing Corporations. Given the sizeable investment required to create grain storage infrastructure, it would be in the interest of the CWC/SWCs that a wider participation of lenders is permitted.

- Restrictions on land holdings: The State Land Holding Laws prohibit consolidation of land holdings.

Financing to farmers with larger land holdings, and thus larger scale of operations, could result in higher credit flows from banks. Further, the Government could also consider consolidation of wastelands and leasing of these to corporates. This would not only enable increase in financing to agriculture, but also aid crop diversification.

EXPORTS OF AGRICULTURAL AND FOOD PRODUCTS

India has 1.5 % (INR 360 bn in 2003-04) share of global agricultural exports (approx USD 522 bn), despite its production leadership in agriculture. India's exports primarily constitutes commodity and primary processed items, where price realisations are low. In addition, many products are showing single digit or negative growth.

The reasons for India's insignificant share in global trade include supply side factors such as lack of consistency in supply and quality, lack of cost competitiveness, and demand side factors such as non-tariff barriers, short product life cycles and perception of Indian food products in overseas markets. Most exporters from India lack scale-for example the largest fresh produce exporter records annual sales of about INR 500mn. This has resulted in lack of economies in operations and renders them uncompetitive. Hence, exporters are not able to establish themselves as long-term players in the export market, rely on opportunistic businesses and are consequently, unable to develop technical and managerial expertise. These factors cumulatively lead to low investments by exporters in brand building, quality improvement and brand development.

FINANCING OF THE FOOD PROCESSING SECTOR

The structural complexion of the Indian supply chain has led to limited scale of financing as well as higher risk, given the lack of control of each of the players, on the supply chain.

Financing of farmers: Farmers often rely on unorganized sources of credit due to bottlenecks in access, timeliness in availability and adequacy of credit from organized sources. The key hurdles faced by banks in financing farmers are their inability to provide adequate collateral as security, and the potential for default, in the absence of an assured market for their produce.

Financing of food processing enterprises: The food processing enterprises primarily comprise small and medium sized companies, a large proportion of which have stand alone operations, with no linkages with farmers, and reliant on other organizations to undertake marketing/further processing of their products. Consequently, companies in the food processing sector usually bear a steep cost of interest for the high risk perception associated with the nature of their operations.

Further there are several regulations which limit availability of finance to the sector, including the definition of priority sector lending, the Cooperatives Act and the Warehousing Corporation Act.

Absence of well-developed risk mitigation tools has further impacted availability of finance to the food and agriculture sector. Additionally, crop insurance schemes in India have faced various issues including assessment of farmer yields at the mandal level, lack of past statistical data to calculate premia, high premia for certain crops and administrative hurdles in managing claims, thus resulting in poor economic viability as also limited benefit to farmers.

FDI in food processing: FDI in the food processing sector is low, constituting a mere 4% of total FDI in the period 1991-2004.The actual inflow is only about 28% of FDI approved (INR 4.2 bn out of approved INR 11.6 bn) This is despite the fact that with the exception of food retailing, plantations and alcoholic beverages, 100% FDI is permitted in the sector.

Government Schemes: Government provides assistance to the industry and entrepreneurs under various schemes. While there are significant overlaps in these on the one hand, there are several unaddressed need gaps of the industry, on the other. Further, the restriction on quantum of financial assistance per unit disincentivises scale. Additionally, the structure of the scheme needs to be based on specific sectoral requirements. The Government needs to have a robust monitoring mechanism to assess the direct and indirect impact of such assistance.

MARKET INTELLIGENCE

Unlike in several leading food producing countries in the world, there is no updated and reliable source of information pertaining to the food processing sector such as on government policies, demand supply outlook, food standards etc. The process of obtaining information from existing industry agencies/bodies, even if available, is tedious. Further, there is limited sharing of information across institutes/bodies collating data. Small and medium sized players are particularly at a disadvantage on account of lack of availability of information.

FOOD SAFETY, HYGIENE AND REGULATIONS

Food safety is a growing concern across the world. There is an increasing need to provide greater assurance about the safety and quality of food to consumers. Compliance with international food standards is a prerequisite to gain a higher share of world trade. At the same time there is growing consciousness among Indian consumers given recent controversies on quality standards of many food products.

The existing mechanism for setting and harmonizing food standards is inadequate in the context of the increasing importance of food hygiene and safety and evolution of international standards. This is due to limited infrastructure for generating/analysing scientific data and food testing. Many laboratories do not have basic facilities to test antibiotic residues, heavy metal contamination and other toxic contaminants in food products. Testing manuals do not describe parameters and procedures adequately. Moreover, limited coordination between the various food testing laboratories has led to inefficient utilization of the food testing infrastructure available in the country.

Further, Indian food regulations comprise various food laws which are under the ambit of various ministries, which has lead to operational complexities for processors.

TAXATION

The tax incidence (excise/customs, state taxes) in Indian food products is higher than many other countries. This has led to low affordability of processed products. Concessions have been extended to different products of food processing industry in the last few budgets but it happened in a selective manner. The reasons based on which concessions were extended for one category hold true for all sectors of the industry.

Apart from the above, there are several other sector-specific issues, which influence the growth of the food sector. These include lack of cultivation of varieties amenable to processing, unequal advantage to the unorganized sector, absence of direct farmer-processor linkages and lack of focus on tapping specific sectoral advantages.

RESEARCH AND DEVELOPMENT

The Indian food industry needs world class technologies to compete effectively with international products. Indian R&D institutions have not been able to develop innovative products, processes and machinery of global stature as reflected in India's share in global trade. The industry is dissatisfied with the quality of research output and response of existing R&D institutions. The key reasons for this are segregation of academics from applied research, inadequate industry interface, low commercial orientation and lack of collaborative efforts with global peers.

AGRICULTURE AND ECONOMIC DEVELOPMENT

As a country develops economically, the relative importance of agriculture declines. The primary reason for this was shown by the 19th-century German statistician Ernst Engel, who discovered that as incomes increase the proportion of income spent on food declines.

For example, if a family's income were to increase by 100 percent, the amount it would spend on food might increase by 60 percent; if formerly its expenditures on food had been 50 percent of its budget, after the increase they would amount to only 40 percent of its budget. It follows from this that, as incomes increase, a smaller fraction of the total resources of society is required to produce the amount of food demanded by the population.

ADVANCEMENT IN FARMING

This fact would have surprised most economists of the early 19th century, who feared that the limited supply of land in the populated areas of Europe would determine that continent's ability to feed its growing population. Their fear was based on the so-called law of diminishing returns: that under given conditions an increase in the amount of labour and capital applied to a fixed amount of land results in a less than proportional increase in the output of food. This principle is a valid one, but what the classical economists could not foresee was the extent to which the state of the arts and the methods of production would change.

Some of the changes occurred in agriculture; others occurred in other sectors of the economy but had a major effect on the supply of food. In looking back upon the history of the more developed countries, one can see that agriculture has played an important part in the process of their enrichment. For one thing, if growth is to occur, agriculture must be able to produce a surplus of food to maintain the growing non-agricultural labour force. Since food is more essential for life than are the services provided by merchants or bankers or factories, an economy cannot shift to such activities unless food is available for barter or sale in sufficient quantities to support those engaged in them.

Unless food can be obtained through international trade, a country does not normally develop industrially until its farm areas can supply its towns with food in exchange for the products of their factories. Economic growth also requires a growing labour force. In an agricultural country most of the workers needed must come from the rural population. Thus agriculture must not only supply a surplus of food for the towns, but it must also be able to produce the increased amount of food with a relatively smaller labour force. It may do so by substituting animal power for human power or by gradually introducing labour-saving machinery. Agriculture may also be a source of the capital needed for industrial growth to the extent that it provides a surplus that may be converted into the funds needed to purchase industrial equipment or to build roads and provide public services. For these reasons a country seeking to develop its economy may be well advised to give a significant priority to agriculture. Experience in the developing countries has shown that agriculture can be made much more productive with the proper investment in irrigation systems, research, fertilizers, insecticides, and herbicides.

Fortunately, many advances in applied science do not require massive amounts of capital, although it may be necessary to expand marketing and transportation facilities so that farm output can be brought to the entire population. One difficulty in giving priority to agriculture is that most of the increase in farm output and most of the income gains are concentrated in certain regions rather than extending throughout the country. The remaining farmers are not able to produce more and actually suffer a disadvantage as farm prices decline. There is no easy answer to this problem, but developing countries need to be aware of it; economic progress is consistent with lingering backwardness, as can be seen in parts of southern Italy or in the Appalachian area of the United States.

PEASANT AGRICULTURE

One characteristic of undeveloped peasant agriculture is its self-sufficiency. Farm families in those circumstances consume a substantial part of what they produce. While some of their output may be sold in the market, their total production is generally not much larger than what is needed for the maintenance of the family. Not only is productivity per worker low under these conditions but yields per unit of land are also low. Even where the land was originally fertile, the fertility is likely to have been depleted by decades of continuous cropping. The available manures are not sufficient, and the farmers cannot afford to purchase them elsewhere.

Peasant agriculture is often said to be characterized by inertia. The peasant farmer is likely to be illiterate, suspicious of outsiders, and reluctant to try new methods; food patterns remain unchanged for decades or even centuries. Evidence, however, suggests that the apparent inertia may be simply the result of a lack of alternatives. If there is nothing better to change to, there is little point in changing. Moreover, the self-sufficient farmer is bound to want to minimize his risks; since a crop failure can mean starvation in many parts of the world, farmers have been reluctant to adopt new methods if doing so would expose them to greater risks of failure. The increased use worldwide of high-yielding varieties of rice and wheat since the 1960s has shown that farmers are willing and able to adopt new crops and farming methods when their superiority is demonstrated. These high-yielding varieties, however, require increased outlays for fertilizer, as well as expanded facilities for storage and distribution, and many developing countries are unable to afford such expenditures.

THE LABOUR FORCE

As economic growth proceeds, a large proportion of the farm labour force must shift from agriculture into other pursuits. This fundamental shift in the labour force is made possible, of course, by an enormous increase in output per worker as agriculture becomes modernized. This increase in output stems from various factors. Where land is plentiful the output per worker is likely to be higher because it is possible to employ more fertilizer and machinery per worker.

LAND, OUTPUT, AND YIELDS

Only a small fraction of the world's land area—about one-tenth—may be considered arable, if arable land is defined as land planted to crops. Less than one-fourth of the world's land area is in permanent meadows and pastures. The remainder is either in forests or is not being used for agricultural purposes.

GENERAL RELATIONSHIPS

There are great differences in the amount of arable land per person in the various regions of the world. The greatest amount of arable land per capita is in Oceania; the least is in China. No direct relationship exists between the amount of arable land per capita and the level of income; Europe has almost as little arable land per capita as Asia and less than Africa; Japan and the Netherlands have very limited amounts of arable land per capita.

The relationship between land, population, and farm production is a complex one. In traditional agriculture, where methods of production have changed little over a long period of time, production is largely determined by the quality and quantity of land available and the number of people working on the land. Until the early years of the 20th century, most of the world's increase in crop production came either from an increase in land under cultivation or from an increase in the amount of labour used per unit of land. This generally involved a shift to crops that would yield more per unit of land and required more labour for their cultivation.

Wheat, rye, and millet require less labour per unit of land and per unit of food output than do rice, potatoes, or corn (maize), but generally the latter yield more food per unit of land. Thus, as population density increased, the latter groups of crops tended to be substituted for the former. This did not hold true in Europe, where wheat, rye, and millet expanded at the expense of pasture land; but these crops yielded more food per acre than did the livestock that they displaced.

As agriculture becomes modernized, its dependence upon land as well as upon human labour decreases. Animal power and machinery are substituted for human labour; mechanical power then replaces animal power. The substitution of mechanical power for animal power also reduces the need for land. The increased use of fertilizer as modernization occurs also acts as a substitute for both land and labour; the same is true of herbicides and insecticides. By making it possible to produce more per unit of land and per hour of work, less land and labour are required for a given amount of output.

RECENT TRENDS

Crop yields have increased dramatically since 1950, with a faster rate of growth in the developing than in the developed countries. Most of this increased output has been due to gains in yields rather than to the expansion of cultivated land. In Europe as well as in North and Central America, the total area under crops has declined; in South America it has increased by more than one-half and in Asia by more than one-third. The large increase in Oceania was due to immigration. The large decrease in Africa was due to a succession of droughts from the 1970s on. Grain yields in the developed regions of the world have increased consistently over the past several decades. In the rest of the world the pre-World War II yields were not achieved again until the mid-1950s. The increases in grain production were more than twice as high in the developing as in the developed countries.

Food production and total agricultural production exhibit nearly identical trends, and changes in food production can be taken therefore as indicative of changes in total agricultural production. Food supplies per capita in developing countries have increased at nearly the same rate as in developed countries, indicating a narrowing gap between food supplies and population growth in the developing countries.

AGROECOSYSTEM STRUCTURE AND FUNCTION

Another distinction to keep in mind is that between agroecosystem structure and agroecosystem function. Agroecosystem structure is how the agroecosystem is organized. It is a consequence of both an agricultural technology system and the environmental and social setting in which the technology is applied.

Agroecosystem structure includes all elements of the ecosystem and how they are connected functionally to one another: *i.e.* all species of crops, livestock, weeds, pests, soil animals, and decomposer organisms-as well as all other plants, animals or microorganisms that are present.

It includes details of soil status and everything about inputs that shape the agroecosystem-the annual calendar of human activities in the fields, sources of labour (*e.g.* family labour or hired labourers), how much capital and energy (*e.g.* petroleum or beasts of burden) are employed, and where they come from (*e.g.* bank loans).

Agroecosystem function is a consequence of agroecosystem structure. Agroecosystem function consists of (a) movements of materials, energy and information from one part of the agroecosystem to another and (b) movements of materials, energy, and information in and out of the agroecosystem. Materials that leave the agroecosystem for human use are regarded as products. We refer to the quantity of these products as production, and system properties concerning production are the ones that customarily have received attention in SUAN research.

EVALUATION OF SYSTEM PROPERTIES

Some complications in assessing agroecosystem properties. It explains how each of the system properties for assessing production has a multiplicity of meanings. This is primarily because there are so many dimensions to production, but it is also because the properties can be very different at different levels of an agroecosystem hierarchy and under different environmental and social conditions.

MULTIDIMENSIONAL CHARACTER OF PRODUCTIVITY

We may wish for a property like productivity to be simple and unequivocal, but in fact it is highly multidimensional because agroecosystems have a variety of products for a variety of uses. Different measures-are more informative with regard to different functions of the products:

- Volume for building materials;
- Biomass (*i.e.* weight) for plant residues or animal manure to be used as organic fertilizer;
- Energy for wood or plant residues to be used as fuel;
- Energy, vitamins, minerals, and amino acids for food;
- Monetary value for exchange purposes.

Incidental outputs of agroecosystems, such as sediment in the water runoff from a field, can also be regarded as 'products', and each of these outputs has its own appropriate measure.

The significance of these different measures is that production of a single agroecosystem may be relatively high for one measure and relatively low for another. A clove plantation is high for monetary value, but low in biomass and food value. A taro field is relatively high in its production of some food values (*e.g.* energy), only moderate in other food values, and generally low in monetary value. Comparison of the production of different agroecosystems is therefore meaningful only when the unit of production is explicitly defined. Monetary value is the most universal measure of agroecosystem production, but no single measure-not even monetary value-is of universal significance.

An equally significant source of the multidimensional character of productivity is that productivity is only meaningful when expressed as production per unit of input. Inputs take a variety of forms-land, labour, energy, cash-and a single agroecosystem's productivity can be quite different with respect to each of the inputs. In general, productivity is high with respect to inputs that limit production and low with respect to inputs that are in excess. For example, where human population density is high, landholdings are small, and labour inputs are intensive (as in high-yielding wet rice cultivation), productivity per unit of land tends to be high but productivity per unit of labour is low. The opposite tends to be true for shifting cultivation in forests with a small human population.

Table. Some Major Sources of Multidimensionality in Agroecosystem Productivity

Measures of Production	Inputs
Biomass	Land area
Food value	Labour
Energy	Materials
Monetary value	Energy
	Cash

The same is true when energy or cash inputs are intensified. In agriculture with heavy cash investments for modern inputs (*e.g.*, fertilizers or pesticides), productivity per unit of land is high, but the net return on cash investments is low compared to traditional homegardens, for example, where cash inputs are much lower.

It may be desirable to express the efficiency of production with respect to very specific inputs. For example: production per unit of water input may be a primary concern in irrigation systems; production per unit of mineral nutrient input may be the major concern with regard to fertilizer costs or where soil nutrient depletion is a problem; production per unit of soil erosion may be the concern if loss of topsoil is imminent; petroleum energy inputs may be more important (particularly to a national government) than other energy inputs if petroleum is imported and therefore consumes precious foreign credits.

RELATIONS BETWEEN AGROECOSYSTEM PROPERTIES

Persons responsible for cropping systems design and other aspects of agricultural development may be particularly interested in the trade-offs between different agroecosystem properties. Improvements in one system property (*e.g.* productivity or stability) should not be at the expense of other properties-or at least the cost should not be too great. It would be convenient to have some simple and general rules to serve as guidelines for how agroecosystems function in this regard, such as 'If productivity increases, sustainability declines', but it is not easy to discern a pattern.

Some highly productive agricultural technology systems are quite stable while others are not, and some low-productivity agricultural technology systems are stable while others are not. For example, intensive high-yield rice production has reliable yields in some areas but is not reliable in other areas because of pests such as the brown planthopper. Consistent relationships between productivity and the other system properties are equally elusive, but exploring those relationships can nonetheless provide some insights into agroecosystem design.

PRODUCTIVITY, STABILITY AND SUSTAINABILITY

There are numerous ways that high levels of productivity can have a positive impact on stability and sustainability. For example, higher productivity may be attained by increasing the harvests in bad years (*i.e.* irrigation to reduce the impact of drought, or pesticides to reduce the impact of pest attacks), thereby making harvests more even from year to year, increasing stability. Higher productivity can be associated with higher sustainability when a more productive crop provides a more complete cover for soil protection and contributes more crop residues for the maintenance of soil organic matter. Higher productivity can also be associated with higher stability or sustainability if it leads to household savings that give a household the capacity to deal with periodic problems that threaten production. In general, any attributes that increase 'fallbacks' and other adaptive mechanisms in an agroecosystem can increase both its stability and/or sustainability.

There are also many ways that productivity can be negatively associated with stability or sustainability.

For example, higher productivity can be associated with lower stability if the higher production is achieved by means of high-yielding varieties that are more vulnerable than local varieties to fluctuating environmental stresses such as droughts and pest attacks-or if high yields lead to a glut on the market that depresses prices.

Higher productivity can be associated with lower sustainability if production is at the expense of soil resources (*e.g.* by generating erosion, reducing soil organic matter or exporting soil nutrients), if the production is due to heavy inputs leading to major alterations in the ecosystem that eventually undercut production (*e.g.* irrigation leading to salinization or pesticides leading to the loss of natural enemies and the emergence of secondary pests), or if higher production is a consequence of labour inputs that place a strain on social institutions underlying the organization of agricultural production.

Higher stability can reduce sustainability in the face of occasional, severe stresses (*i.e.* reduce resilience) if, under stable conditions, the agroecosystem (and its inhabitants) cease to exercise their abilities to deal with stress (because there is no need to do so) and consequently lose that ability, even though they may eventually need it. Farmers with a steady supply of irrigation water have more stable yields than rainfed agriculturalists because they are liberated from the negative effects that short periods without rainfall can have on rainfed agriculture. However, they may also lose the agricultural technology they once had for rainfed agriculture, simply because they no longer need it. Drought-resistant varieties may be discarded and cultivation practices to make the most of limited soil moisture supplies may be forgotten. As a consequence, they may not have the means to prevent crop failure if the irrigation system should fail. There are numerous other examples of this conflict between stability and resilience.

Chemical fertilizers help to buffer farmers from spatial variations in soil quality in their fields. Because large amounts of labour are required to collect and transport animal manure or green manure to maintain organic matter levels, the effort may not seem necessary as long as chemical fertilizers can compensate for diminishing organic matter. However, the impact of an increase in fertilizer prices that forces farmers to reduce fertilizer use can be particularly severe if they have not taken the effort to maintain the organic matter content of their soil.

To cite another example, the construction of flood control dams allows farmers to cultivate fertile flood plains without worrying about flood damage, but the 'once in a hundred years' flood that overruns the dams can cause damage on a scale far greater than would occur if the farmers pursued their agriculture and constructed their villages in constant expectation of floods. In pest control, the use of chemical pesticides can increase stability, providing an opportunity to eliminate even the smallest pest losses. Indigenous pest-resistant crop varieties may be discarded, and the pesticides may eradicate the pests' natural enemies along with the pests. If a pesticide-resistant strain of the pest should suddenly appear, the damage may be more serious than it would have been without pesticides, because natural enemies are no longer present to keep the pest abundance within reasonable bounds. Even if the development of pesticide resistance is gradual, it eventually may be necessary to increase the frequency of pesticide application to a point where the crop must be discontinued due to excessive pesticide costs.

ECONOMICS OF VEGETABLE PRODUCTION

Tomato is the most important vegetable crop in the area. It was grown by 95.3 percent sample households and occupy 12.73 percent of total crop area. Share of tomato in total area under vegetable crops was about 40 percent. Size category wise input use, output and return from vegetable crops are studied by taking tomato as the representative crop.

IRRIGATED FARMS

Tomato cultivation at irrigated farms shows that there was a mixed pattern in use of inputs related to farm size in various panchayats. Use of farm yard manure, fertilizers and family and total labour was higher at holdings above 2 acres compared to the holdings below 2 acres. Bullock labour use was higher at marginal and small size holdings compared to other holdings. Expenditure on plant protection material was highest at the other farms in Dharot panchayat and at marginal farms in Deothi panchayat.

CONDITIONS IN VARIOUS PANCHAYATS IN SOLAN BLOCK, 1992-93 PARTICULARS DHAROT DEOTHI BHOJNAGAR

Yield, total value of output and net return per acre were higher at the other farms in Deothi and Dharot, while in Bhojnagar small farms showed better performance compared to the other size categories. Input use and yield in Deothi and Dharot panchayats which have access to road was significantly higher compared to the sample farms in Bhojnagar located away from market and roadside. Net return from tomato production was many times higher in the villages located around roadside compared to the distant villages. Farm gate price realised by Deothi and Dharot farmers was 1.5 to 2.25 times the price of tomato obtained by Bhojnagar farmers.

Demand for offseason vegetables like tomato produced in Himachal Pradesh comes from outside states. Vegetable growers in Deothi and Dharot which are located around pucca road enjoy the advantage of transporting their marketable produce to markets in Delhi, Punjab and Haryana which fetches attractive price. The second reason for high price obtained by Deothi and Dharot farmers is the adjustment in crop growing season to get output in most tomato scarce period. Deothi farmers, who are in the business of vegetable production since long time, have almost perfected the technique of producing vegetables in offseason. On the other hand, Bhojnagar farmers have to carry the saleable output to a distance of 1-6 km involving steep slope; this causes heavy cost in terms of labour time and the net farm gate price reduces to quite low level. At the roadhead, the produce is purchased by private traders who operate there for few days when there is enough marketed output. Irrigated area in Bhojnagar is small and the output from this area is neither sufficient to attract private traders to buy it at the nearest roadhead nor it is economic for the producers to transport the produce to market. Therefore, irrigated farms also have to synchronise their crop production with the vegetable production in rainfed area-on which vegetables are produced in monsoon season. Thus the advantage of high price for tomato in early season cannot be availed by the irrigated farms in the areas not having easy access to road or market.

Input use as indicated by cost A, and use of family labour in irrigated tomato crop was inversely related to farm size. Yield, value of output and net return per unit of area were significantly higher for marginal and small farms than for other farms. This shows that the production performance under irrigated conditions at the land holdings of the size below 2 acres is better compared to the other holdings.

UNIRRIGATED FARMS

Effect of infrastructure on productivity and profitability was quite sharp under unirrigated production also. Per acre value of output (TVP) of tomato in Kotho panchayat was Rs. 5 to 8 thousand higher compared to Bhojnagar, though yields were higher in Bhojnagar. The reason was that Kotho panchayat was located nearby the agricultural market in Solan town due to which Kotho farmers realised 50 to 120 percent higher farm gate price compared to Bhojnagar farmers.

Based on performance of tomato crop it can inferred that size of farm is not a constraint to diversify through offseason vegetable cultivation. Rather, under irrigated conditions performance of lower sized farms was better compared to higher sized farms. As we have observed earlier, there were large variations in crop intensity across farm size categories. In such situations production performance is better indicated by aggregate of all crops per unit of net cultivated area.

CROP ECONOMICS PER UNIT OF NET CULTIVATED AREA

Economics of aggregate crop production per unit of net cultivated area according to size classes is present. The estimates based on net cultivated area are better indicator of economic performance of aggregate crop production, as they take into account the difference in performance due to crop intensity. Per acre use of purchased as well as home produced inputs was highest at marginal farms and lowest at farms having more than 2 acres of cultivated holdings in all the panchayats. Even the use of hired labour was highest at marginal farms. The most important factor underlying this kind of trend was percent of total crop area put under vegetables cultivation, which showed inverse relation with farm size But the hard working farmers used to bring water in buckets from distant places to meet water requirement of vegetable crops during the stress period. In this way, the marginal farmers offset to some extent the disadvantage they have with respect to irrigation, and put higher percent of area under vegetable crops as compared to other categories. Furthermore, marginal farmers in Dharot, Kotho and Bhojnagar made best efforts to overcome the irrigation constraint by using more human labour, higher quantity of other inputs and more use of bullock labour to prepare fine soil structure. As a result of these factors, value of total output per unit area was highest at marginal farms and lowest at other categories, except one case in Dharot where small farms recorded somewhat higher TVP than marginal farms.

In Deothi, where marginal farmers have advantage over other categories in access to irrigation, TVP per acre of net sown area for them was 30 percent higher than for small farmers and 63 percent higher than the TVP at other farms having holdings above 2 acres. The increase in TVP per unit of NCA over TVP per unit of TCA, was highest in the case of 'marginal' farm category and lowest in the case of 'other' category. The increase in TVP due to crop intensity at marginal farms varied between 63 to 93 percent in various panchayats.

The net return over cost Al was Rs. 12,827, 14,976 and 13,016 per acre of NCA at marginal, small and other farms in Dharot and Rs. 17,872, 14,054 and 11,557, respectively, in Deothi. The net return in Bhojnagar and Kotho for the three farm size categories was Rs.3151, 4,076, 3,134 and Rs.5,692, 4,985 and 6,912 for the respective categories.

In Dharot, marketed surplus as percent of total output was 89 percent for marginal and 87 percent for small and other farm size category, in Dharot. In Deothi, marginal and other farmers sold 86 percent and small farmers sold 84 percent produce. In Bhojnagar, marketed surplus ranged between 52 to 69 percent, the highest being at marginal farm size. In Kotho, sales comprise 81 percent of production at marginal farms, 78 percent on small and 84 percent on other farm category. This shows that marketed surplus per unit of area at marginal farms was either higher or same as at the bigger farm size categories in most cases, though the common Indian experience is that proportion of marketed surplus in total production is either near to zero or very small at smaller size farms, and it increases with the increase in farm size.

Our findings show that if there is, incentive, marginal farms can supply same or even higher percent of production to the market as the large farmers. In our sample, incentive was provided by crop diversification through vegetable crops which are produced with purely commercial considerations for sale in the market. Since the marginal farmers allocate higher proportion of cultivated area under vegetables, the share of vegetables in total output was also higher, which resulted in higher percent of marketed surplus in output compared to other farm size categories.

SALE PATTERN OF VEGETABLES

The sample farms disposed off their marketed surplus of vegetable crops at many places namely village roadside market, nearest agricultural market, Delhi market and various markets in Punjab and Haryana state. Bhojnagar farmers who neither have easy access to road nor to any market sold their entire surplus at the nearest roadside market irrespective of the size of the holdings. Sample farms in Kotho, which were located at a distance of 1-4 km. from the local market and 1-3 km. from road preferred to sell most of their produce in local market. Size wise, marginal and small farmers ventured to sell about 2 percent produce in Delhi market whereas other farmers sold 20 percent produce in Delhi Market.

Deothi farmers sold 83-90 percent produce in markets outside the state and remaining quantity in Solan market. About 78-81 percent produce from Dhrot panchayat was sold in Delhi and the neighbouring states. In Dharot and Deothi, small and marginal farmers sold only slightly higher proportion of their produce in roadside and local market compared to other farms. Size of holding in these two panchayats did not impose any constraint to market output to distant markets.

The sales pattern of vegetable crops in various panchayats show that it was not size of holding but access to infrastructure like road and market which determine marketing behaviour or the vegetable producers. In the areas where vegetables were produced by large number of growers and there was easy access to transport, small lot does not cause any problem to transport produce to remunerative markets. Where there are large number of producers the option to share common truck for transporting the produce is very easily available and forthcoming.

INCOME AND EMPLOYMENT IMPLICATIONS OF DIVERSIFICATION THROUGH VEGETABLES

The estimates of labour use in vegetables and other crops, for aggregate of all panchayats, were used to study the employment implication of shift in crop pattern at block and district level. The calculations reveal that one percent shift in area from other crops to offseason vegetables would lead to 1.20 and 1.60 percent growth in existing level of labour employment depending on whether the shift takes place in irrigated or unirrigated area. The average of the two would be close to 1.60 percent due to higher share of unirrigated area in total crop area.

The employment growth would be higher in areas other than Solan block because area under vegetable cultivation in Solan block was highest among all blocks of the district. When the estimates were extended to district level, it was found that one percent shift in irrigated area from other crops to vegetable crops would result in 1.60 percent growth in existing level of employment. The increase for unirrigated area worked out to be 2.85 percent. These estimates show the potential of diversification through vegetable crops on 'onfarm' employment opportunities. Another dimension of labour use in vegetable cultivation is that it reduces seasonality in labour use because the peak time of labour requirement of vegetable crops differ from the peak labour demand in most of the other crops.

The impact of shift in crops pattern, in favour of vegetables, on net income was quite sharp. The net return from vegetable cultivation was 2 to 23 times higher than other crops under irrigated conditions and 3 to 40 times higher under unirrigated conditions. A one percent shift in area from other crops to vegetable crops in a district like Solan was estimated to lead to around 6 percent increase in net return from existing cropping pattern under irrigated conditions and around 4 percent increase when the shift in area takes place on unirrigated land.

POLICY IMPLICATIONS

Agricultural diversification through vegetable crops has a huge potential for employment and income generation in Western Himalayan Region. Vegetable cultivation, due to its labour intensive nature, is more beneficial for marginal and submarginal holdings where family labour availability per unit of land is higher compared to bigger size holdings. This is the reason that percent of area shifted to vegetable crops increase with the decrease in farm size in the vegetable growing area. Our findings based on micro level investigations reveal that in case of commercial and higher profitability enterprises, farm size is not a constraint for production and marketing.

The study shows that it is not correct to assume that marginal and small farmers do not have sufficient land to put under commercial crops after allocating land to foodgrains to meet the family needs. It was found that where economic incentive is available, farmers allocate area based on relative profitability irrespective of the foodgrains requirement of family which can be easily met through purchases. There is a strong evidence that it is not the farm size, but infrastructure like access to motorable road, market and irrigation which determine the extent, success and profitability of diversification through high paying crops like offseason vegetables. Promotion of enterprises like offseason vegetables would go a long way in generating productive employment and income in the hill areas in Western Himalayan Region where size of holdings and per capita arable land are very small and traditional crops with low productivity are not capable of providing sufficient income and employment to the population dependent on agriculture sector.

CLIMATE CHANGE ON CROP PRODUCTION IN INDIA

Climate is one of the main determinants of agricultural production Through out the world there is significant concern about the effects of climate change and its variability on agricultural production. Researchers and administrators are concerned with the potential damages and benefits that may arise in future from climate change impacts on agriculture, since these will affect domestic and international policies, trading pattern, resource use, and food security.

The researchers are of the opinion that while crops would respond favourably to elevated CO2 in the absence of climate change, the associated impacts of high temperatures, changed pattern of precipitation and possibly increased frequency of extreme events such as drought and floods, would possibly combine to reduce yields and increase risks in agricultural production in several parts of the globe.

In India agricultural production is often determined by the whims of nature. The climate Marginal and Small Farms change is expected to result in higher temperatures and rainfall. The higher expected temperature might lower the yields. However, at the same time, higher rainfall could enhance growing period of crops. Also the higher concentration of CO2 in the atmosphere under changed climatic conditions might act as aerial fertilizer and enhance crop yields.

All these factors have to be taken in to consideration while examining the climate change impact on agriculture. During the 1990s, researchers repeatedly claimed that global warming would have dire consequences for key crops. Professor Richer Adams, an agricultural economist says, "If you just take an agronomic model and make conditions hotter and drier, then, yes, crop yields go down." " But if you are a farmer, you see your crops aren't doing so well and plant a more resistant type. In India, climate change is expected to make an impact in agriculture, resulting in lower yields of crops. The objectives of this paper are to examine the effect of fertilizer and labour inputs on crop productivity and to investigate the impact of climatic variables such as rainfall and temperature on crop production. Climate response functions have been estimated, using regression model by incorporating weather variables, in order to examine the impact of climate change on productivity of two crops. The impact of weather variables has been examined and the study is confined to two states.

METHODOLOGY

In a study Kaufmann R. K. and Seth E Snell (1997) had specified yield as a linear function of purchased inputs. For the purpose of analysis seed, fertilizer and labour inputs were considered in the regression model as explanatory variables and yield as the dependent variable. However, crop yield is affected not only by these purchased inputs but also by climatic variables and social factors.

This integration allows complementing the earlier methodologies, which concentrate only on the purchased inputs and thereby, allows to better evaluating the adaptation strategy to climate change. The results show that climatic variables account for 19% of the yield change while the social variables account for 74% of the yield change.

Kavi Kumar and Parikh (2001) has also established a functional relationship between farm level net revenue and climatic variables, with a view to estimate the climate sensitivity of crop production in Indian agriculture.

To assess the impact of climate related variables and agricultural production in India, the present study has been undertaken. For this purpose, two crops *viz.* Rice and Jowar have been selected. These two crops are predominantly grown in monsoon season and any change in climate, particularly rainfall and temperature would effect the productivity of these crops significantly. In the present paper an econometric bio-model of crop production has been attempted. Agricultural production depends on not only climate related variables, but also on use of several factors like fertilizer, labour and other resources. For the purpose of present study, the following model has been used:

Y= f (F, HL, AR, NR, DFNR, MAXTEMP, MINTEMP)

Where,

Y= Crop yield on per hectare basis

F = Fertilizer used/ha.

HL = Human labour in hours/ha.

AR = Actual Rainfall (mm)

NR = Normal Rainfall (mm)

DFNR = Deviation from normal rainfall (mm. or %)

MAXTEMP = Mean maximum temperature during crop season (0C)

MINTEMP = Mean minimum temperature during crop season (0C)

The model was estimated for Rice and Jowar crops. The analysis was undertaken for the country as a whole, using state wise data for both these crops. The study was also undertaken using district wise data as well. For Rice crop, the state of Orissa has been selected, because this is the main crop of the state. For Jowar, the state of Karnataka has been selected, because of it's significance in the region. Data on different variables for various states has been collected from the publications of Directorate of Economics and Statistics, Ministry of Agriculture, as well as from the Economics and Statistics Directorates of various states.

Similarly district-wise data has been primarily collected from the publications of these two state govts. Least square technique has been used to estimate different regression equations for the country as a whole as well as for both the states. Several forms of equations were analysed and results were computed. The analysis was done under usual assumptions. It has been hypothesized that productivity of the selected crops is positively influenced by the application of chemical fertilizers and human labour. These two are the main purchased inputs in the production of these crops. Climate, in the present model, has been approximated by rainfall – actual, normal as well as deviation from the normal. Other variables representing the climatic factors are mean maximum and mean minimum temperature in crop growing period. This has been further hypothesized that extreme variations in these variables such as excessive rain or scanty rain and very high or low temperature would adversely affect the crop productivity in the selected states and districts for the selected crops.

RESULTS AND DISCUSSION

In order to study the impact of climate change on productivity, regression equations were estimated using the model and data stipulated in the methodology. As already mentioned, earlier various forms of functions were estimated and out of these, linear form provided the best fit and therefore, only linear regression equations have been presented for Rice and Jowar crops for the country as well as both the selected states. For rice productivity and climate, using state-wise data for the country as a whole. Variable pertaining to fertilizer use, human labour, actual rainfall and temperature pertains to year 2000-01, whereas normal temperature consists of observations from a period of 30 years. Variation in the productivity of Rice. This implies that productivity of Rice on per hectare basis is influenced by these variables. Moreover, the fertilizer use variable had a highly significant coefficient, suggesting a significant impact of fertilizer use on productivity.

Surprisingly, the variables pertaining to actual rainfall and maximum temperature, during the growing period of the crop, had negative coefficient, implying that these two variables had negative impact on productivity. The mean minimum temperature had a statistically significant and positive coefficient, implying that this variable would effect the productivity of Rice positively.

Estimated regression equations for Rice crop and climate related variables, using data from various districts of Orissa state. Similar to the earlier analysis, results pertaining to the linear regression have been presented.

The results indicate that the variable of fertilizer use, normal rainfall, minimum and maximum temperature could explain 28% of variation in Rice yield. When actual rainfall variable was replaced by the deviation in the rainfall from the normal, the value of R2 dropped down to 7 percent, implying that the actual rainfall had better explanatory power, than that of deviation in rainfall from normal. Results also indicated that mean maximum and minimum temperature could explain a larger part of variation in the Rice yield. Maximum temperature had a negative coefficient. It has implication that the increasing temperature would effect the productivity of Rice negatively, in the state.

The results indicated that the use of fertilizer variable has a negative impact on Jowar productivity, which is not correct on theoretical grounds. However, the human labour variable has a positive and statistically significant impact on Jowar productivity. The deviation from normal rainfall had negative impact on productivity. This implies that rainfall contributes significantly in raising productivity of Jowar, which is normally a rainfed crop. The mean maximum temperature, during the growing season of the crop had a negative implication for the crop, while the minimum temperature, during the same period, would effect the crop productivity positively. The R2 in all five selected equations are good and it's value ranges between 40% to 52%.

The rainfall during the crop growth period has also negative coefficient. Whereas mean maximum temperature will effect the crop positively. The results indicate that the effect of climate related variables have mixed effects on productivity of Jowar, as rainfall shows negative impact, while maximum temperature has a positive effect. However, for understanding the real effect of these variables on productivity of Jowar crop, a detailed study is required.

Climate plays an important role in shaping agricultural production in India. Lack of irrigation makes agriculture a gamble with nature. The effects of climatic variability is quite visible in case of majority of farmers who are marginal and small and lack resources required for adjustment for climatic variations. The present study covers rice and jowar crops for the country as a whole as well as for the states of Orissa and Karnataka.

The study reveals that excessive rains and extreme variation in temperature would affect the productivity of these crops adversely thereby affecting the incomes of farming families in a negative manner. Thus suitable strategies pertaining to resource use, planting flood and drought resistant varieties of crops, better irrigation networks, and crop mix are to be adopted for mitigating the harmful effects of climatic changes.

RICE RESEARCH AND DEVELOPMENT: IMPACT ON PRODUCTION

The development of the modern semi-dwarf rice varieties and the use of increased inputs of fertilizer, irrigation, and crop protectants ushered in the first phase of the Green Revolution in rice growing countries (Fischer and Cordova 1998, Swaminathan 1999). Loans made available for development of irrigation infrastructure and conducive government policies led to widespread adoption of the new technology. The impact on food supply was immediate. This followed the development of seed multiplication infrastructure and seed distribution by extension systems, training of rice scientists and extension personnel in the way of the new generation of rice production technologies. With this also began the collection and conservation of rice genetic resources and their evaluation and utilization in rice breeding programmes.

The second phase of green revolution emphasised on stabilising yields by incorporating resistance to multiple pests and diseases and also reducing the growing period of rice cultivars. Other developments included introduction of farming systems methodology and mechanization of land preparation and threshing. The international sharing and testing of rice germplasm was also initiated and formalized.

These concerted efforts led to increased productivity, selfsufficiency, and even surpluses in many countries of South and Southeast Asia. In Asia as a whole, rough rice production shot up from 240 M tonnes in 1956 to 513 M tonnes in 1996, an increase of over 114 per cent over 30 years. This increase in production, which surpassed a population increase of 82 per cent in Asia during the same period, resulted in increased per capita availability of rice. The average rice yields in Asia increased from 2.1 t/ha in 1966 to 3.8 t/ha in 1996, an increase of ca. 2 per cent per year. The green revolution in rice was most dramatic and most successful in irrigated rice ecosystem. But large parts of South and Southeast Asia, where rainfed lowlands are the predominant rice ecosystems, have less benefited from rice revolution.

Consequently, the yield growth has remained slow, at 1.4 per cent per year, much below the population growth rate. More than three decades have passed since the advent of green revolution, and there are some concerns that the potential of the green revolution technology might have already been exhausted. A new green revolution is required so that at the prevailing price levels, the increased demand for rice by 50 per cent to 60 per cent over the next 30 years could be achieved. Accordingly, the new research targets focus on increasing yield per hectare by capturing the yield potential of existing cultivars and pushing the yield frontier up with tropical hybrids and new plant type, enhancing input use efficiency by increasing nutrient and water use efficiency, reducing labour costs, implementing integrated pest management, and sustaining resource base in the irrigated ecosystem.

Modern tools and procedures of genomics, cell culture, molecular biology, genetic engineering which, compared with traditional breeding methods, are more precise and rapid with high turnover are being deployed for new breakthroughs in yield potential. Significant strides have been made in hybrid rice research and hybrid rice farming has become a ground reality in India. The hybrid rice technology confers a 15 per cent yield advantage over the best inbred varieties.

The technology is also spreading in Bangladesh. Efforts are underway to enhance the heteroic level of the tropical hybrids to make hybrid rice cultivation more profitable. Possible development of apomictic hybrids, which would avoid the need for F1 seed production, would make the technology even handy for the poor resource farmers. Similarly, the prototypes of the new plant type are already at the field-testing stage and are likely to reach farmers' fields soon. These genotypes promise about 20 to 25 per cent increase in yields (Khush 1995). Linking heterosis to still more desirable traits, such as insect and disease resistance or tolerance to abiotic stresses, such as salinity and alkalinity, can help the crop perform even better and substantially offset the cost of production.

For impact on rainfed rice, the research targets are focusing on developing tolerance to abiotic stresses, maintaining natural resilience of ecosystem to pests, supporting direct seeding methods for crop establishment, incorporating micro-nutrients in rice, and improving grain quality. If the rice research could succeed in incorporating traits that help modern varieties withstand abiotic stresses and maintain high yield, then such varieties will be adopted readily and more extensively in the unfavourable ecosystem because the risk in cultivation will be reduced. Farmers will have the incentive to apply chemical fertilizers at optimum levels, which will further improve rice yields.

Modern biotechnology tools can enhance these efforts by increasing the efficiency of conventional breeding and by using genetic engineering to bring non-rice or alien genes into the rice gene pool (Bennett 1995), enabling selection of genetically complex traits with high precision and monitoring of changing virulence in pests. Some of the promising research programmes in progress are engineering resistance to stem borer, leaffolder, bacterial leaf blight, sheath blight, and rice tungro virus, and improving nutritive quality by increasing ßcarotene (vitamin A) and protein content using novel genes from near and distantly related life forms. However, the commercial application of biotechnology products in general may still be somewhat problematic.

A major policy issue is the regulatory climate governing the release of genetically engineered products or generically modified crops to ensure public health and environmental integrity. The countries in South Asia, including India, have yet to equip with necessary regulatory mechanisms to import, test, and use such materials. Intellectual property protection and plant breeder's right may further delay easy access to materials developed through the use of biotechnology tools.

Although concerted efforts to improve rice production in rainfed ecosystems began only about a decade ago, the progress is quite encouraging. Two major developments that took place in early 1990s, greatly spurred research of rainfed ecosystems. IRRI organized the research programme according to the rice ecosystems that ensured adequate support and opportunities for rainfed rice.

The rice research consortia that also began operating early 90s, brought together the resources from NARS in South Asia and IRRI to solve the problems of individual rice ecosystem in a concerted and systematic way by sharing resources and responsibilities. Participatory sharing of responsibilities for conducting strategic research under the Rainfed Lowland Research Consortium. The results are already there to show the impact of these complementary efforts.

With lager and better trained scientific staff, deeper and broader knowledge of rice production problems and constraints, increased emphasis on applied as well as strategic research issues, and greater motivation of scientists, the Asian NARS are contributing substantially to the development of technologies that will increase and sustain higher production of rice in their own countries and in other rice growing countries of the world.

SOME EMERGING CONCERNS

By 2025, 50 per cent to 60 per cent more rice needs to be produced. However, the land, labour, and water, the basic resources responsible for production are getting increasingly scarce and fatigued and are in need of efficiency gain. Pollution of soil, water, air, bio-mass, and production of greenhouse gases is also being linked to inefficient use of agro-chemicals and agri-inputs. There is thus an urgent need for a thorough assessment of the opportunities for increasing land productivity, including crop diversification. Similarly, as the opportunity cost of family labour rises, given alternative employment opportunities, there is need for labour-saving technologies. Growing water scarcity for irrigated rice ecosystem is another serious challenge for which, while technological options, such as recycling, exist, but little has been done in respect of complementary policies. At present, more than 5 tonnes of water is needed to produce a kilogram of rice. The irrigated rice environments not only face increased competition for water, but also face reduced water supplies due to system degradation and reduced investments in infrastructure development. In South Asia, the growth rate in irrigated area has dropped from 2.8 per cent during 1975-80 to 0.1 per cent during 1985-88 period (Rosegrant and Swendsen 1992). It is absolutely necessary that technologies that are promoted must sustain rather than degrade the long-term productivity of the resource base.

Since the beginning of 1990s, unit production costs are beginning to rise and rice farmers are facing declining profits. The stagnant yield frontiers and diminishing returns to further intensification are the primary reasons for the current reversal in rice profitability. Further, rapid withdrawal of labour, division of land to other uses, increased competition for water, and withdrawal of subsidies on agricultural inputs have all led to increasing the cost of rice production. In the absence of shift in yield frontiers for rice and/or a substantial improvement in input use efficiencies, the profitability of rice production in Asia may revert to the pre-green revolution era.

The South Asian countries made a great leap in food production over the last three decades. The technological innovations complemented by public investments and policies that promoted intensive rice production systems helped bring about the change in food supply. However, there is a growing sense of complacency about the current and future food supplies in this region. Lack of concern is seen in declining national budgetary allocation for maintaining existing rice infrastructure, including irrigation, reduced bilateral and multilateral donor support for expanding irrigation, a deliberate policy shift towards diversification out of rice, and a substantial cutback in research funding at both national and international levels. Emphasis has shifted to natural resource management and environmental sustainability. However, there are signs that this complacency is not warranted, as greater challenges for producing extra rice in future exist.

Although currently the countries of South Asia are pursuing the policies for achieving self-sufficiency in rice production, it is anticipated that over the next two decades, some of them may have to go for rice import to make up for further domestic shortfalls. Pingali *et al.* (1997) have indicated that domestic shortfalls could occur due to: (i) continued and unabated population growth, (ii) rapid economic growth in some countries that will make the maintenance of self-sufficiency unprofitable, and (iii) trade liberalization and agreements that could lead to Asian food sector becoming more open to international competition. While the short-term effects of GATT agreement could be modest, the longterm effects could be substantial (Pingali 1995). Small countries that are self-sufficient now in rice, may find it profitable to import at least a part of their rice requirements in exchange of diverging production resources to more remunerative activities. However, for countries with large rice-eating population and low-income, such as India and Bangladesh, achieving food security through imports may not be possible due to high volume involved and foreign exchange constraints.

For these countries, increasing productivity on rice lands should remain a top priority. Among the technological options, hybrid rice is a commercially viable technology with 15 per cent yield advantage over the best inbred varieties. India, first among the countries outside China to develop and commercialise hybrid rice technology, however, has yet to take full advantage at the desired level. As against one million hectares targeted for 2000, not even one-fifth could so far be planted to hybrid rice (Siddiq 2002). Inability to meet the seed demand, a factor that contributed to slow pace of adoption, would have to be overcome.

Also, hybrid seed quality would have to be ensured to avoid inconsistent yield performance. Free exchange of rice germplasm provided unlimited resources for rice improvement programmes and sharing of improved materials among the rice growing countries. However, due to imposition of the intellectual property rights and other WTO requirements, the restriction on germplasm exchange is fast growing.

Under these circumstances, countries with poor genetic resource base and not-so-well developed research set-up, such as Nepal and Bhutan, would face difficulties in accessing improved rice germplasm. The countries in South Asia could, however, develop some collaborative mechanisms to ensure exchange and use of rice germplasm among themselves without affecting each other's interest and sovereignty right. Adequate increase in budgetary support for rice research and development is long overdue in countries in South and Southeast Asia so that it becomes comparable to that in developed countries.

The seed-fertilizer technology of green revolution era was simple and its impact was easily visible. The new technology which deals with complexity of the natural resource management, input use efficiency, and sustainability needs clear understanding and better skills for its transfer and adoption. Its impact also is slow, but with long-term effects. The challenge for the research system is in finding efficient and cost-effective methods for transferring knowledge-intensive crop management technologies to farmers. The ultimate goal is to have farmers who are capable of deriving the appropriate recommendations for each farmer's field using 'if-the' type analysis. The current practice of providing blanket recommedations of fertilizer application to farmers over large areas is no longer a valid mode of operation for extension workers. To reduce knowledge gap, the extension services in future would have to transfer knowledge and provide decision-making options rather than straight input use recommendations. The extension technicians and farmers, therefore would have to be well trained, and the scientists more responsive to convince farmers of the importance of the technologies and how they work. Information technology could play an important role in rice production and natural resource management. It would require upgrading of educational curricula of agricultural schools and colleges. India, being a leader in the field of information technology, could gain substantially by introducing precision farming approaches at farmer level.

The application of modern technologies, such as biotechnology, remote sensing, meteorological information for medium and long-term forecasts, crop modelling, geographic information system, etc., requires trained personnel and state of the art research facilities. Except, perhaps, India, few other countries in South Asia have the required capacity to absorb and utilize these technologies for the benefit of rice production.

Working closely with advanced laboratories or through the International Rice Research Institute could therefore be advantageous. It would be detrimental to agricultural economy as a whole, if these countries are left behind in harnessing the benefits of these advanced technologies in solving some of the present problems. Also substantial policy reforms ought to be in place before the fruits of biotechnology would become available to the rice farmers in South Asia. A holistic food security policy need not be anti-environment; it can actually contribute to long-term environmental sustainability. Rational food production systems that improve the efficacy of inputs and resource use, help conserve the natural resource base rather than threaten it.

CONTENT FOR COMMERCIALISATION OF SMALL FARM AGRICULTURE

The crux of agricultural growth in India is to maintain a rising level of foodgrain production and a reduction in rural poverty by speedy growth of income and employment. The new seed-fertiliser based technology has helped the farmers achieve the first goal of food security, because priority for foodgrain production is rightly compatible with subsistence orientation of the majority of the farmers-especially marginal and small farmers. In the past, it was believed and in fact verified by several studies that the new technology is scale neutral and the poor would share at least regularly with rich in the gains from the technology.

This has not happened and poverty persists. It is a paradox that a record foodgrain production of 186 million metric tonnes in 1994-95 coexists with nearly a third of the country's population living below the poverty line. Indeed, the economic inequalities have widened over the years, revealing that integrating capitalist agriculture with welfare goals is more formidable than earlier thought of. One reason is the lack of adequate attention in policies and programmes for agricultural growth to the need for rapid commercialization of agriculture. The history of the presently developed economies shows that modernisation, monetisation and integration are the essential ways of agricultural growth.

The new technology helped modernization. Further, the increasing demand for money-to purchase most of the input off-farm, leads to increasing sensitivity of farmers to relative prices of agricultural commodities and that helped monetization and commercialization. The next step is to ensure the transfer of resources (generated surpluses)-wage goods, raw materials and investible funds-from agricultural growth and improve the capacity of fanners to adopt new agricultural technology.

Therefore, it would increase the income of the farmers and contribute to creating employment for the landless. This is the process and theory of 'Integrated Rural Growth'. The farm size, direction and pace of this transfer are important. Indian experience is that the process has by-passed a vast majority of the producers in agricultural sector and the progress in integration is highly inadequate. While modernisation and monetization have helped agricultural growth-largely in foodgrain production and commercialization on a limited scale, the process has failed to encompass the marginal and small farmers. The paper analyses the causes and consequences of this deficiency in integration.

MARGINAL AND SMALL FARMS

By their numerical strength, the marginal and small farmers form the majority of cultivators in the country as a whole and in Tamil Nadu and Madurai district which form the universe for the present study. The pattern of size distribution of farms is similar-the predominance of marginal and small farms in both number and area operated by them. Compared to all India, their predominance is much greater for Tamil Nadu and Madurai district. This served the purpose of this study-namely to understand the scope and constraints for commercialization of the small farms and to identify the policy options therefore:

DATA

The study is necessarily based on primary data, because the required information was not available with any source of secondary data. Therefore, a sample of 200 farms was selected in Madurai district by a three stage random sampling method-the stages being CD blocks, villages and farms and by distributing the sample among the villages in probability proportion to the number of farms in each village to the total number of farms in all the 20 sample villages in five C.D. blocks. These sample farms were post-stratified into three size groups based on this operational area *viz.*, small (< = 2 ha); medium (2-4 ha) and large (>4 ha). By definition therefore, small group included marginal farms and the large group included medium farms and the group of semi-medium in the secondary data corresponded to the medium size group of the sample.

CROPPING PATTERN

The concept of commercialization refers to the increasing share of the marketable surplus in the total farm business income of the farms. This share can be increased by (i) generation of more marketable surplus in subsistence oriented foodgrain crops; (ii) by increasing production of market oriented cash crops and other products (collectively known as high value adding enterprises) and (iii) both. An increase in foodgrain production through improvement in productivity of land and diverting area for the high value adding (HVA) crops or other enterprises will ensure food security and commercialization simultaneously.

Therefore the cropping pattern of the farms will show the share of high value adding crops. Groundnut, onion, chillies, sugarcane and cotton are the value adding (cash) crops cultivated by the farmers of the district, while paddy, jowar and blackgram are the food crops raised by them.

In the total gross cropped area of 297.05 ha, cash crops accounted for 64.03 ha (21.56 percent) only showing the farmer's preference to food crops-a sign of subsistence orientation. The share of the food crops in gross cropped area was as high as 90 percent in small farms as compared to 59.75 percent in large farms.

Therefore commercialization through production of products for market (cash crops) was very small *i.e.*, less than 10 percent in total area under cash crops. The large farms were just 12 (six percent) in the sample of 200 farms, but their share in area under the cash crops was 53.60 percent. Therefore the commercialization of agriculture would need attention to small farms, to encourage area under cash crops.

PRODUCTIVITY OF CROPS

The average productivity (in kg/ha) of the crops raised in the sample farms. In jowar, bajra, blackgram and groundnut (rainfed crop), the small farms had the highest productivity. In other crops excepting sugarcane, the productivity of small farms was not significantly different from that in other groups. Only in sugarcane average productivity per hectare was just 101 tonnes/ha as compared to 140 tonnes/ha on large farms. Therefore, productivity was not a serious constraint for commercialisation of small farms if the marketed surplus were not different for the size groups. By the time of enquiry (April '95) the sample farmers had completed the sale of crops raised in the previous years. Therefore their marketed surplus was estimated and expressed as percentage to total production.

It is readily seen that small farmers had the lowest percentage of marketed surplus in food crops, with only exception of blackgram, but they had 90 percent of GCA under them. The area under cash crops was just 10 percent but their marketed surplus was 100 percent with one exception of onion where it was 98.70 percent. The inference is that production in small farms is more subsistence oriented than market oriented. This is a constraint for their commercialization.

FARM INCOME

The income of the farm included income from crops (sale proceeds plus imputed value of produce retained by the family), livestock, off-farm income from wages earned by the members of the family going for work in other farms and income from other sources such as non-farm employment and property income. In all the size groups, income from crops exceeded sixty percent, but it was least on small farms. Income from livestock was negligible (0.92 percent) on small farms, about 15 percent on large farms and 21.42 percent on medium farms.

The small farms would be able to maintain livestock with the help of straw from grain crops that dominated the crop pattern, but they had not, largely because members of the family went out to work on other farms as shown by the large (35.74 percent) wage income from such works. Large farms had very little family labour and therefore, their attention was mostly on crops and cows and poultry were maintained mostly to meet the home needs.

It was the farms in the medium size group that had family labour which preferred maintaining livestock rather than going to work off-farm. Hence, income from livestock had a share of 21.42 percent of gross income, while off-farm wage income was just 1.40 percent. Other sources of income included non-farm employment of family members, properties and other miscellaneous sources. The share of this source increased that too very significantly as the farm size increased. Small farms were seen to be better than the medium farms in their operational efficiency as shown by the gross income/cost ratio and net income. Then operational ability of the farmers is not a constraint for their commercialization.

LIVESTOCK

The farmers maintained one or two cows/she-buffaloes, work bullocks, sheep and poultry birds. Income from livestock supplemented the income from crops. But the size of livestock depended upon the availability of family labour to attend them. As members of the family of small farms worked as wage earners off-farm and non-farm, the livestock maintenance was on a small scale and brought very little income to the farms. It was not a commercial activity. This can be seen in the source-wise income of the farms.

COMMERCIALIZATION

In most of the recent literature, commercialization is approximated by the relative share of cash crops in the GCA. A more accurate measure of commercialization of farming (as opposed to commercialization of crop production)-especially in the context of a diversified agriculture including livestock and allied activities-is the percentage share of income from the sale of farm products in the total value of production. What did not enter the market would be retained for use on-farm and its value had to be imputed.

The percentage value of farm products sold out to the total value of production was just 29.44 percent on small farms, while the value of retained products accounted for 70.56 percent, thus clearly revealing that small farms were subsistence oriented. In contrast, the share of sales in total value product was 61 percent on medium farms and 82 percent on large farms showing them to be largely market oriented. What would explain this situation? Opinion of the farmers was analysed and it revealed three major causes for their reservation on expanding area under cash crops and lack of interest in cultivation of high value adding horticultural crops such as vegetable, fruit and flower crops.

In seasonal cash crops which included groundnut, cotton, sugarcane and chillies, the non-availability of credit was the most important constraint, followed by risk and unfavourable price. In livestock products of dairy, poultry and piggery, non-availability of creditand want of buyers (*i.e.*, non availability of market outlet within their easy reach) were the reasons for limiting the size of those enterprises. In horticultural crops which included vegetables, fruits and onion, the need for making huge investments with production lag and the risk involved due to their bulkiness and perishability and also absence of market were the reasons reported by the farmers, as the constraints. The inability of the farmers to invest, reflected on the internal credit rationing, while their complaint of non-availability of credit would show the external credit rationing. Therefore, credit rationing and unfavourable market emerged to be the two major limiting factors for the commercialization of small farms. Interaction with officials of the commercial banks and Primary Agricultural credit Banks showed that they considered small farms as unviable units even for short term lending, not to speak of long term growth credit. This statement was verified by analysing the ability of the farmers to repay the credit by comparing their income and expenditure levels and possible savings per annum.

PROBLEM FOCUS

On an average the family of a small farm spent Rs. 10,380 per year, while it was Rs. 21,242 for medium farms and Rs 65,516 for the large farms. At this level of expenditure, there was little scope to reduce the consumption expenditure of a small farm family, which is neatly one half of the medium size group and one sixth of large farms. When this level of expenditure was compared with average annual income of the farms, small farm had only a negative saving, while medium and large farms saved nearly 10 percent and 19 percent of their income respectively.

So, bankers are right that small farms were not viable for credit. Only way for making them viable was to increase their income. It should be further emphasised that any increase in income would be spent to increase consumption first rather than saving, because the marginal propensity to consume is 0.82 (estimated from the sample data). Therefore income would have to raise much more than what was required to remove the presently seen negative savings. To allow for atleast five percent savings, the average income of small farms should raise to Rs. 11,500.

This was feasible with change in enterprise mix in favour of cash crops and improving productivity of the crops-both additional investment for which the average small farmer was not credit worthy. The problem is made more serious by two more facts. First majority among small farmers had farms smaller in size than the mean size, and thus they are problem farmers. Secondly, there was wide variation in net income from crops and the coefficient of variation of the net farm income was 18.10 percent. This might bring more farmers within the risk net and credit would have a low level of probability of recovery. Therefore a low income-low investment cycle exists and it sustains rural poverty. How to break this cycle and help a cumulative growth path to small farmers?

POLICY OPTIONS

One way to break the cycle is to make credit available on the basis of prospective income or project based lending. The results of this study clearly show that commercialization of farming significantly contributes to the net farm income. Size of the farm was not a constraint for it; only capital rationing and market size were the real constraint. The crop pattern seen in medium farms are technically feasible and economically viable. Therefore alternate crop patterns may be identified and evaluated for their economic viability with the help of either partial budgets or linear programming technique to maximize aggregate net income of the farms. In the exercise, horticultural crops and allied activities may be the decision variables to be included. With the estimated costs and returns, plans with required level of savings may be identified and commanded to the bankers. The advantage of this method is that the identified plans will win the confidence of both farmers and the lenders. Its major limitation is the cost and time involved in the exercise.

A second option is to encourage collective action such as group farming or cooperative farming, wherein farmers pool in their resources voluntarily to achieve larger scale of operation. Even medium size group farms were the net positive savers.

Therefore, such a collective action is feasible. This is the approach suggested by Parthasarathy when he argues for alleviating poverty by a process of institutional reforms. Past experiences in institutional reforms such as cooperatives, regulated markets and C.D. blocks place a heavy discount on this strategy. However, institutional reforms in the past were attempts to impose institutions on the farmers who never developed a sense of partnership in them. This weakness may be overcome by educating and guiding farmers to organise themselves rather than inviting them to join a sponsored institution. With this new approach, it is likely that cooperatives may emerge to be a practical way, of course with much of the present bureaucratic control discarded. It is of course a long term prospect.

A third option is the marketisation of the farm sector, with continued subsidy on inputs to the marginal and small farmers. This is consistent with the New Economic Policy which places faith on liberalization and invisible hand. However, the basic condition for such a policy to succeed is the presence of perfectly or near perfectly competitive conditions in the market. Such a condition does not exist and it will take a very long time to integrate farmers into the system.

The required process of adjustment will ruthlessly eliminate poor and less organised and institutional cannibalism may be inevitable and that makes the process both painful and anti-thesis of liberalization for social uplift of the weaker sections-one being the small farmer. Therefore only a combination of the above options will succeed. It must be recognised that as a social group, farmers are averse to collective action, unless the provocation is very strong. They also resist any imposition of institutions on them. However, if they have to survive in a liberalised economy, there is no alternative for an effective organisation for collective action. Only recently there is some sign of lobbying among farmers.

Exploiting this trend, they should be helped to evolve their own organisation. This will be the real remedy to the small farmers' problems. Yet, they have to be helped with project based lending for joint borrowing by a group of small farmers.

AGRICULTURAL ADVISORY SERVICES AND THE MARKET

Promoting market orientation in agricultural advisory services aims to provide for the sustainable enhancement of the capabilities of the rural poor to enable them to benefit from agricultural markets and help them to adapt to factors which impact upon these.

As a study by the Overseas Development Institute demonstrates, a value chain approach to advisory services indicates that the range of clients serviced should go beyond farmers to include input providers, producers, producer organisations and processors and traders.

MARKET INFORMATION

Efficient market information can be shown to have positive benefits for farmers and traders. Up-to-date information on prices and other market factors enables farmers to negotiate with traders and also facilitates spatial distribution of products from rural areas to towns and between markets. Most governments in developing countries have tried to provide market information services to farmers, but these have tended to experience problems of sustainability. Moreover, even when they function, the service provided is often insufficient to allow commercial decisions to be made because of time lags between data collection and dissemination.

Modern communications technologies open up the possibility for market information services to improve information delivery through SMS on cell phones and the rapid growth of FM radio stations in many developing countries offers the possibility of more localised information services. In the longer run, the internet may become an effective way of delivering information to farmers. However, problems associated with the cost and accuracy of data collection still remain to be addressed.

Even when they have access to market information, farmers often require assistance in interpreting that information. For example, the market price quoted on the radio may refer to a wholesale selling price and farmers may have difficulty in translating this into a realistic price at their local assembly market.

Various attempts have been made in developing countries to introduce commercial market information services but these have largely been targeted at traders, commercial farmers or exporters. It is not easy to see how small, poor farmers can generate sufficient income for a commercial service to be profitable although in India a new service introduced by Thompson Reuters was reportedly used by over 100,000 farmers in its first year of operation. Esoko in West Africa attempts to subsidize the cost of such services to farmers by charging access to a more advanced feature set of mobile-based tools to businesses.

MARKET INFRASTRUCTURE

Efficient marketing infrastructure such as wholesale, retail and assembly markets and storage facilities is essential for cost-effective marketing, to minimise post-harvest losses and to reduce health risks. Markets play an important role in rural development, income generation, food security, developing rural-market linkages and gender issues. Planners need to be aware of how to design markets that meet a community's social and economic needs and how to choose a suitable site for a new market. In many cases sites are chosen that are inappropriate and result in under-use or even no use of the infrastructure constructed.

It is also not sufficient just to build a market: attention needs to be paid to how that market will be managed, operated and maintained. In most cases, where market improvements were only aimed at infrastructure upgrading and did not guarantee maintenance and management, most failed within a few years. Rural assembly markets are located in production areas and primarily serve as places where farmers can meet with traders to sell their products. These may be occasional (perhaps weekly) markets, such as haat bazaars in India and Nepal, or permanent.

Terminal wholesale markets are located in major metropolitan areas, where produce is finally channelled to consumers through trade between wholesalers and retailers, caterers, etc. The characteristics of wholesale markets have changed considerably as retailing changes in response to urban growth, the increasing role of supermarkets and increased consumer spending capacity. These changes require responses in the way in which traditional wholesale markets are organized and managed.

Retail marketing systems in western countries have broadly evolved from traditional street markets through to the modern hypermarket or out-of-town shopping centre. Despite the growth of supermarkets there remains considerable scope to improve agricultural marketing in developing countries by constructing new retail markets. However, there is little point in undertaking market development improvements unless they result in a positive socio-economic impact. Effective regulation of markets is essential. Inside the market, both hygiene rules and revenue collection activities have to be enforced. Of equal importance, however, is the maintenance of order outside the market. Licensed traders in a market will not be willing to cooperate in raising standards if they face competition from unlicensed operators outside who do not pay any of the costs involved in providing a proper service.

MARKETING TRAINING

Farmers frequently consider marketing as being their major problem. However, while they are able to identify such problems as poor prices, lack of transport and high post-harvest losses, they are often poorly equipped to identify potential solutions.

Successful marketing requires learning new skills, new techniques and new ways of obtaining information. Extension officers working with ministries of agriculture or NGOs are often well-trained in horticultural production techniques but usually lack knowledge of marketing or post-harvest handling. Ways of helping them develop their knowledge of these areas, in order to be better able to advise farmers about market-oriented horticulture, need to be explored. While there is a range of generic guides and other training materials available from FAO and others, these should ideally be tailored to national circumstances to have maximum effect.

RECENT DEVELOPMENTS

New marketing linkages between agribusiness, large retailers and farmers are gradually being developed, *e.g.* through contract farming, group marketing and other forms of collective action. Donors and NGOs are paying increasing attention to ways of promoting direct linkages between farmers and buyers. The growth of supermarkets, particularly in Latin America and East and South East Asia, is having a significant impact on marketing channels for horticultural, dairy and livestock products Nevertheless, "spot" markets will continue to be important for many years, necessitating attention to infrastructure improvement such as for retail and wholesale markets.

ENABLING ENVIRONMENTS

Agricultural marketing needs to be conducted within a supportive policy, legal, institutional, macro-economic, infrastructural and bureaucratic environment. Traders and others cannot make investments in a climate of arbitrary government policy changes, such as those that restrict imports and exports or internal produce movement. Those in business cannot function if their trading activities are hampered by excessive bureaucracy and form filling. Inappropriate law can distort and reduce the efficiency of the market, increase the costs of doing business and retard the development of a competitive private sector. Poor support institutions, such as agricultural extension services, municipalities that operate markets inefficiently and export promotion bodies, can be particularly damaging. Poor roads increase the cost of doing business, reduce payments to farmers and increase prices to consumers. Finally, the ever-present problem of corruption can seriously impact on agricultural marketing efficiency in many countries.

PRODUCTION AND PRODUCTIVITY GROWTH IN AGRICULTURE

When evaluating the performance of a production unit or the agricultural sector, it is common to use production (the level of output), productivity (output per unit of input) or efficiency (actual output relative to the potential output or best practices) as indicators. Although these measures are closely related, they can yield different rankings in measuring performance. In general, productivity is the most commonly used measure, be it measured in terms of total factor productivity (TFP) or in partial terms such as labour productivity (output per labour) and yield (output per hectare) for its relative ease in calculation and interpretation.

The changes over time in output and productivity growth in different regions are compared and the causes for variations are discussed.

AGRICULTURAL DEVELOPMENT AND INPUT USE

Agricultural output and productivity vary greatly with the stage of economic development, resource endowments, government policy and agronomic-ecological conditions. However, there is a similar path in agricultural development over time and across countries. Pingali and Heisey (1996) categorized the technological transformation of cereal crop production system into three distinct phases:

- The land-augmentation phase;
- The labour-substitution phase; and
- The knowledge-and management-intensity phase.

The basic assumption is that the transition from one phase to another is determined by growing factor scarcity, first for land, then for labour and finally for other factors of production, such as machinery and management skills.

The first phase is characterized by area expansion being the main source of output growth, as was seen during the pre-Green Revolution era of the 1950s and 1960s. However, as opportunities for area expansion decline over time, cropping intensity is increased, along with increasing use of water, fertilizers, pesticides and high yielding varieties. This was indeed the case during the Green-Revolution period in the 1970s and early 1980s. Such intensive production results in an increased demand for labour and mechanization, as the production system moves from single-cropping to double-and triple-cropping with increased application of purchased inputs. Eventually, production reaches the point of diminishing marginal returns to further intensification, as was the case in the late 1980s, the post-Green Revolution phase.

Here, better technical knowledge and management skills are used to substitute for traditional inputs. Variety selection, fertilizer timing and placement, water management and pesticide application are some areas in which productivity has improved with reduction in unit cost of production. The model just outlined is used in the following analysis as the basic framework to explain the changes in input use and in productivity between the 1960s and the 1990s. First, it is applied to various regions in the world, then to the developing and developed countries, and finally to the selected countries in Asia and the Pacific.

FACTORS AFFECTING PRODUCTIVITY GROWTH

In explaining productivity growth, economists originally limited themselves to the role of conventional inputs such as land, labour, physical capital, water and chemical inputs. However, the failure to explain productivity growth adequately led them to examine the role of human capital and public goods, such as education, agricultural research and extension and publicly provided infrastructure. Public policies that have a strong link to agricultural productivity such as policy reforms were also examined.

The rationale for considering research is the belief that investments in research result in increases in the stock of knowledge, which, in turn, either facilitate the use of existing knowledge or generate new technology. Technological advances, whether resulting from changes in input quality or how inputs are combined, lead to productivity gains. Education, training and extension also increase productivity by increasing people's knowledge and skill base, which are essential for technology adoption and efficient use of inputs. Public infrastructure, on the other hand, increases productivity by facilitating the exchange of goods and services.

AGRICULTURAL RESEARCH AND EXTENSION

Many researchers have explored the roles of research and extension in promoting agricultural growth. Rosegrant and Evenson (1992) found that in South Asia, public research accounted for 30 percent of the output growth, and extension for about 25 percent, with corresponding rates of return being 63 percent and 52 percent, respectively. Pray and Evenson's (1991) survey of Asia found the rates of return to national research investment ranged from 19 to 218 percent, returns to national extension investment from 15 to 215 percent, and returns to international research investment from 68 to 108 percent.

Evenson and McKinsey (1991) found that public investment in research accounted for over half of the output growth in India and extension contributed about one-third. The calculated internal rates of return were 218 percent for public research and 177 percent for extension. However, they found that little output growth was attributable to infrastructure. Jamison and Lau (1982) also found that physical capital had little impact on production or profits, as compared to farmer's education and extension services.

Fan (1996) found that public research expenditures accounted for about 20 percent of total production growth in Chinese agriculture during the period 1965 to 1994. The annual rates of return to agricultural research investment in China ranged from 44 percent to 83 percent.

Fan (1996) concluded that the rapid growth in agricultural output in China during the 1980s and 1990s was the result of public investments in R&D as well as the institutional and market reforms that began in 1979. He concluded that increases in agricultural research were justifiable; not only did they stimulate additional output growth, but the rate of return to agricultural research was much higher than commercial interest rates.

Despite the high rates of returns from public research investments, agricultural research intensity (ARI), measured as a percentage of Chinese agricultural GDP, was found to have declined from 0.56 percent for the period 1958-1965 to 0.43, 0.44, 0.39 and 0.40 percent, respectively, for 1966-1976, 1977-1985, 1986-1990 and 1991-1993 (Fan, 1996). Lin (1998) reported that, as part of the overall market reform, the Chinese Government had reduced its fiscal appropriation for agricultural research, shifting funding from institutional supports to competitive grants and cost recovery. As such, it can be expected that an increasing proportion of research activities will move from the public to the private domain.

Other studies on output growth have also shown a high payoff from agricultural research and extension. The results indicate that the rate of return on research, in most cases, ranged from 15 to 50 percent for both developed and developing countries, but some estimates were as high as 218 percent. The wide disparity among the estimates raises questions regarding the sensitivity of these estimates to the commodity of interest and the use of different time periods and methodologies.

Estimates for Asian countries appear to be higher and show a much wider variation than those of studies in the United States. This could be due to the diverse nature of Asian agriculture, which differs from country to country in economic, social and agronomic-ecological conditions. Because of inconsistency in the data and methodology used, it is not possible to make direct comparisons across countries or over time. Nevertheless, the general conclusion that R&D yields relatively high returns seems indisputable.

TECHNOLOGICAL CHANGE

Technological change is recognized by many as one of the most important sources of productivity growth (Antle and Capalbo, 1988). It refers to the changes in the production process that come about from the application of innovation and newly acquired scientific knowledge and technical and management skills. Technological change increases agricultural productivity either by shifting the production frontier upward so that more measured output can be produced with the same amount of inputs or by moving closer to the production frontier so that the same amount of output can be produced with a smaller amount of inputs. Better organizational and management skills not only improve input-output combinations but enable producers to respond more quickly to changing market circumstances.

While generation of new technology or knowledge comes from investments in research and development, adoption of technology involves investments by the potential users in both physical and human capital. Therefore, adoption of technology depends principally on their applicability and expected returns of the innovation. However, there may be a long lag between development, adoption and productivity gains. Chavas and Cox (1992) found the lag to be up to 15 years between making an investment in research and having an effect on productivity. However, after taking effect, the benefits from an innovation may persist for thirty years or more.

The lag between generation of new technology and its widespread adoption by farmers has important policy implications. First, the adverse effects of reduced public funding to agricultural research and extension on productivity may be under-estimated if the lagged effects are not accounted for. Secondly, the complementarity between research and extension should be taken into account. The former helps the development of new technology, while the latter helps speed up the rate of diffusion and adoption of new technology. Extension can be done more effectively by identifying factors that contribute to technology adoption. As an example, innovators in a farming community can be identified and targeted for extension services.

Since better-educated farmers are found to be more likely to adopt new technology, human capital is a pre-condition for technology adoption and hence productivity growth. Further, if adoption of new technology requires additional investments, lack of access to credit and additional inputs may prevent or slow down technology adoption. Finally, because potential users of new technology often differ in the agronomic-ecological conditions in which they operate, new technology may require adaptive research before it can be transferred successfully to different locations. These impediments to technology adoption mean careful planning and provision of necessary infrastructure are essential to capture the full benefits of new technology.

IMPACT OF ECONOMIC REFORMS ON AGRICULTURE IN INDIA

The serious foreign-exchange crisis in 1990 led to a number of well-publicized economic reforms in the early 1990s dealing with trade, industrial licensing, and privatization. The reforms had an impact on the agricultural sector through the central government's effort to withdraw the fertilizer subsidy and place greater emphasis on agricultural exports. The cut in the fertilizer subsidy was a result of the government's commitment to reduce New Delhi's fiscal deficit by removing grants and subsidies from the budget. The government action led to a reduction in the use of chemical fertilizers and protests by farmers and opposition political parties. The government was forced to continue the subsidies but at a somewhat lower level.

New import and export policies aim at enhancing export capabilities of the agricultural sector by increasing productivity and promoting modernization and competitiveness. The measures to facilitate export growth include allowing the import of capital for the agricultural sector, reducing the list of agricultural products that cannot be exported, and removing the minimum export price from a number of products. Agricultural exports increased by 30 percent in FY 1991 and 14 percent in FY 1992 in terms of rupee value, but declined by 8 percent from FY 1990 to FY 1992 in United States dollar terms because of the devaluation of the rupee in 1991.

In the mid-1990s, it was expected that agriculture would continue to be the most important sector of the economy for the rest of the decade in terms of the proportion of GDP. However, even when it is not the sector providing the largest share of GDP, the importance of agriculture is not likely to diminish because of its critical role in providing food, wage goods, employment, and raw materials to industries.

Despite their preoccupation with industrial development, India's planners and policy makers have had to acknowledge the critical role of agriculture in the early 1990s by changing basic policy. The gains in agricultural production should not lead to complacency, however. Continuing increases in productivity, developing allied activities in rural areas, and building infrastructure in rural areas are essential if India is to continue to be self-reliant in food and agricultural products and provide a modest surplus for exports.

AGRICULTURE AND RURAL DEVELOPMENTS

From a nation dependent on food imports to feed its population, India today is not only self—sufficient in grain production, but also has a substantial reserve.

The progress made by agriculture in the last four decades has been one of the biggest success stories of free India. Agriculture and allied activities constitute the single largest contributor to the Gross Domestic Product, almost 33% of it. Agriculture is the means of livelihood of about two—thirds of the work force in the country. This increase in agricultural production has been brought about by bringing additional area under cultivation, extension of irrigation facilities, the use of improved high yielding variety of seeds, better techniques evolved through agricultural research, water management, and plant protection through judicious use of fertilizers, pesticides and cropping practices.

IRRIGATION

As efforts continued to increase the irrigation potential in the country, the last 40 years saw the gross irrigated area reach 8~ million hectares. Flood forecasting has become an important activity over the years. Over 500 hydrological stations collect and transmit data through 400 wireless stations for issuing forecasts for 157 sites. About 5000 forecasts are issued in a year with 94% accuracy. The country also receives international support, with the World Bank as a primary source, for developing the water resources. International cooperation is also envisaged in setting up a National Centre for Information on Water and Power. As there is a broad seismic belt in the country, particularly along the Himalayan, the Kutch region and ports of Maharashtra, a scheme is being evolved to collect all data on seismic activity at various dam sites.

CROPS

The 1970s saw a multi-fold increase in wheat production that heralded the Green Revolution. In the next decade rice production rose significantly; in 1995-96, rice production was 79.6 million tons. Total grain production crossed 191 million tons in 1994-95, a big leap from 51 million tons in 1950-5 1. During the Seventh Plan, the average grain production was 155 million tons, 17 million tons more than the Sixth Plan average.

To carry improved technologies to farmers, a National Pulse Development Programme, covering 13 states, was launched in 1986. The Special Food Production Programme augmented efforts to boost pulse production further. In 1995-96, pulse production was 13.2 million tons. With some States offering more than the statutory minimum price, sugarcane production also received a boost, in 1995-96 a record 283.0 million tons was registered.

FERTILIZERS

The fertilizer industry in India has grown tremendously in the last 30 years. The Government is keen to see that fertilizer reaches the farmers in the remote and hilly areas. It has been decided to decontrol the prices, distribution and movement of phosphatic and potassic fertilizers. Steps have been taken to ensure an increase in the supply of non-chemical fertilizers at reasonable prices.

There are 53 fertilizer quality control laboratories in the country. Since bio—fertilisers are regarded as an effective, cheap and renewable supplement to chemical fertilizers, the Government is implementing a National Project on Development and Use of Bio-fertilisers. Under this scheme, one national and six regional centres for organizing training, demonstrating programmes and quality testing of bio-fertilisers has been taken up.

It was a challenging decision of the Government to take Bombay High gas through a 1,700-km pipeline to feed fertilizer plants located in the consumption centres of North India.

However, the major policy which has ensured the growth of the fertilizer industry is the thrust on accelerating fertilizer consumption by fixing, on the one hand, low and uniform price for fertilizers, and on the other hand providing the manufacturers adequate compensation through the retention price and subsidy scheme. As expected, fertilizer nutrient demand has gone up from 0.29 million tons in 1960-61 to 13.9 million tons at the end of 1995-96, compared to 12.15MT during 1992-93.

FOOD PROCESSING

A Ministry of Food Processing Industries was established in July 1988 to ensure better utilization of farmers' output by inducting modem technology into the processing of food products, thus augmenting the income of farmers and generating employment opportunities in rural areas. A new seeds policy has been adopted to provide access to high quality seeds and plant material for vegetables, fruit, flowers, oil-seeds and pulses, without in any way compromising quarantine conditions.

Initiatives have been taken to encourage private sector investment in the food processing industry.

FISHERIES

Fish production achieved an all-time high of 4.9 million tons at the end of 1995-96. Programmes that have helped boost production include the National Programme of Developing Fish Seeds, Fish Farmers' Development Agencies and Brackish Water Fish Farmers' Development Agencies. The Central Institute of Fisheries Nautical and Engineering Training trains the necessary manpower. To diversify fishing methods and introduce processed fish products on a semi-commercial scale, an Integrated Fisheries Project has been launched. A National Fisheries Advisory Board has also been established.

AGRICULTURAL RESEARCH

The apex body for education, research and extension education in the field of agriculture is the Indian Council of Agricultural Research (ICAR), established in 1929. India's transformation from a food deficit to a food surplus country is largely due to ICAR's smooth and rapid transfer of farm technology from the laboratory to the land. ICAR discharges its responsibilities through 43 research institutes, four national research bureaus, 20 national research centres, nine project directorates, 70 all-India coordinated research projects, and 109 Krishi

Vigyan Kendras (farm science centres). Besides, the programme of Agricultural Education is coordinated by ICAR with the curricula and other normative guidance given to the 26 Agricultural Universities and four National Research Institutes.

DRINKING WATER

A Technology Mission on Drinking Water and Related Water Management has been constituted to cover the residual problem villages and provide potable water at 40 liter per capita per day, and 70 liters per capita per day in desert areas inclusive of 30 liters for cattle. The Mission is tackling the problem through 55 mini-missions in project districts and countrywide problem oriented sub-missions. A Village Level Operation and Maintenance (VLOM) pump called India Mark-11 has been developed and is being exported to 40 countries. By March 31, 1993, over 79% of the rural and about 85% of the urban population was provided drinking water facilities.

OILSEEDS PRODUCTION

A Technology Mission on Oilseeds was launched in 1986 to increase production of oilseeds in the country and attain self-sufficiency. Pulses were brought under the Technology Mission in 1990. Before the Mission was launched in 1985-86, oilseed production was 10.83 million tons; during 1995-96, it was estimated at 22.42 million tons, which is a record. Soybean, rapeseed and mustard largely contributed the increase in production. Production of pulses has seen many ups and downs, which is expected to be checked under the Mission. The country grows mainly nine oilseeds, with groundnut, rapeseed and mustard accounting for 62% of total production. Lately, soybean and sunflower have shown major growth potential.

AGRICULTURAL GROWTH AND PERFORMANCE

Although India's economic reforms were initiated in June 1991, the process of liberalization was implemented gradually and thus it is difficult to assess the full impact of the liberalization measures. Nevertheless, an attempt is made to discuss what is observable in terms of agricultural growth.

One observation is that the expected increase in exports due to liberalization simply did not occur. India's share in world exports was 0.6 per cent in 1997; India has to aim for at least 4 per cent by 2005 in order to meet the growing import demands for capital goods, raw materials and crude oil as well as to meet her external financial commitments (Kalirajan and others 2001). For the last decade or so, India's share in world exports of agriculture has been between 2 per cent and 3 per cent. India is not as competitive as the other countries and calculations show that India's crop yields have increased at a slower rate over the 1990s.

In addition, the agricultural sector's output growth decreased to 2.9 per cent during 1992-1993 to 1998-1999. Kalirajan and others (2001) explain that two important reasons for the slowdown are that there was no major breakthrough in developing new high-yielding varieties during the 1990s and there was a decline in the environmental quality of land which reduced the marginal productivity of the modern inputs. What could this mean in terms of the effectiveness of the policies of reduced protection to industry, a market determined exchange rate and the opening of the agricultural sector to foreign trade?

First, although the reduction to protection of industry is substantial, there is reason to believe that the reduction was not necessarily sufficient to benefit the agricultural sector whose tariffs were also drastically reduced. Hence, the expected shift in resources to agriculture did not occur. Second, is the apparent ineffectiveness of the market determined exchange rate in boosting exports. This is however not surprising as the exchange rate may not be a key factor determining agricultural export demand for India. In general, unlike manufacturing industries, agriculture did not benefit much from these two policies because the share of imported inputs in the value of agricultural production is small. It is likely that a change in the mindset and attitude of farmers has yet to take place and there are delays or hesitation in embracing India's openness.

Third, in opening up the agricultural sector to foreign trade, India has taken major steps towards trade liberalization since 1991, partly on its own initiative and partly from its commitments to WTO. Kalirajan and others (2001) provide a detailed review of these reform procedures. But why have the benefits from trade liberalization been slow to come?

One reason is that prospects for growth in agricultural exports depend partly on domestic policies and partly on the removal of protectionist policies pursued by developed countries such as Japan and members of the European Union (EU). An OECD report (1998) estimated that the producer equivalent subsidy in the OECD countries increased by US$ 9.3 billion from 1988 to 1993 and this subsidy as a percentage of the value of production in 1997 was 9 per cent in Australia, 20 per cent in Canada, 47 per cent in EU and 70 per cent in Japan.

These protectionist practices do not seem likely to come to an early end. An UNCTAD report (1999) noted that 29 member countries of the OECD spent an average of US$ 350 billion a year in agricultural support between 1996-98. Schumacher (2000) further reports that the EU provides product-specific trade distorting domestic support to at least 50 different agricultural products. The implication of these reports is that food exports from India may not show a large increase given the international environment and the still-existing restrictions on exports in the major importing markets based on the self-sufficiency argument and food security. Other macroeconomic factors, such as the recession in developed countries in 1996-98 as well as the 1997 South-East Asian financial crisis, have clouded the possibilities of increasing Indian exports.

Another problem faced by Indian agricultural exporters is the protectionist measures in the form of non-trade barriers that developed countries use to restrict market access. This is by tightening requirements of quality, testing and labeling, and anti-dumping and countervailing measures. For example, in May 1997, the EU banned marine products from India citing unhygienic processing conditions. The extra costs of meeting the standards required in export markets as well as costs associated with changes in the production mix and transactions associated with exports may well be discouraging Indian exporters.

One existing problem of India's agricultural protection is the use of input subsidies. The general argument favouring this has been that it is necessary to encourage the use of particular inputs for production for various benefits. For India, Gulati and Sharma (1995) show that the input subsidy in per cent of GDP increased from 2.13 in the triennium ending 1982-1983 to 2.73 in the triennium ending 1992-1993. But the benefits of these subsidies have accrued to only certain classes of farmers in some regions cultivating irrigated crops. Furthermore, highly subsidized prices of inputs such as irrigation water and electricity for pump sets have encouraged cultivation of water-intensive crops, over-use of water, ground water depletion/salinity and water logging in many areas. Subsidy for nitrogen fertilizer on the other hand has resulted in nitrogen phosphorous potassium imbalance and acted as a disincentive for use of the environmentally friendly organic manure. As a result, the linkage between food crops and non-food crops, which include fodder, has been reduced.

These adverse consequences are a drain on the fiscal burden of central and state Governments. Thus, if not properly monitored, input subsidies can be counterproductive and, in this context, protection to lower costs of production should be done selectively in the course of liberalization.

In fact, Agenda 21 of the United Nations Conference on Environment and Development in 1992 stressed that there is a need for integration of environmental considerations in the pricing of natural and other resources in such a way that prices reflect social costs.

Such a pricing policy will not only lead to a more efficient use of scarce resources but also result in subsidy reductions and improvements in environmental quality. The money saved from the reduction of subsidies can be spent in the development of rural infrastructures, agricultural research, farmers' education and other forms of support for agriculture.

While Tamil Nadu had the highest yield in rice, oil seeds and sugarcane, Punjab enjoyed the highest yields in wheat, coarse cereals, pulses and food grains. Karnataka on the other hand is seen to do well in cotton and Bihar performed quite well in pulses and coarse cereals. Further analysis and findings by Kalirajan and others (2001) show that Punjab had made remarkable achievements on the agricultural front while Bihar had remained stagnant in the last two decades, with Karnataka and Tamil Nadu showing moderate achievement. Clearly, differences in physical endowments, climatic conditions and institutional characteristics are some of the reasons for the varying productivity performance. Thus, having across the board economic reforms is likely to work less effectively than state-specific policy measures that enable each state's agricultural yields to reach their full potential. The comparative advantage of each state's agricultural production should be determined and with inter-state restrictions removed, total agricultural output would see a very significant increase.

For example, Karnataka with less favourable soil and water resources should be given incentives to concentrate on agro-processed products and corporate agriculture in horticulture, floriculture and animal husbandry, or to undertake watershed development to help with dry land agriculture. Many studies have indicated that with watershed areas, productivity growth has been mainly due to seed and fertilizer use. Thus, this state has to be given input subsidies for high yielding seed varieties but at the same time, the farmers need to be educated on the over use of chemical fertilizers.

With Bihar, agricultural performance is problematic on many fronts. First, although demographic pressure has increased and agricultural technology has improved, most of the uncultivated land is concentrated in southern Bihar, where irrigation facilities have not kept pace and the soil is of poor quality. Given the physiography of southern Bihar, wells are also unsuitable and thus the dominant mode of irrigation has been through tanks whose expansion and maintenance has been neglected. Second, the infrastructural facilities of Bihar have been lagging as seen by the infrastructure development. Due to infrastructural bottlenecks, availability of modern goods and services has not increased or their supply remains costly or unreliable.

Third, agriculture in Bihar is dominated by small and marginal farmers and the prevalence of mass poverty is largely related to the backwardness of agriculture. Fourth and importantly, the state agricultural policies in Bihar are in dire need of review. The semi-feudal production condition still exists in rural areas and the ineffective protection of tenancy rights has hindered agricultural growth. The slow pace of land consolidation reflects inadequate financial outlays and a shortage of manpower. Kalirajan and others (2001) note that marketing and extension services in Bihar are also rather weak compared to the other states.

Punjab on the other hand, was one of the few states which enjoyed the success of land reforms and the high priority of investment in rural infrastructure. Also, the irrigation base of the small and medium sized farms was comparable to that of large farms. In addition, the Punjab Agricultural University at Ludhiana contributed to the development of new seed varieties.

However, there are clear signs of a decline in crop yields since the 1990s and this has been associated with the increasing use of fertilizers and excessive water use which have increased the unit cost of production as a result of declining soil quality. Hence, care is needed when providing further input subsidies in fertilizer and water use. Another related fact is the steep increase in wages in Punjab and in the absence of productivity increases, the cost increase has affected the profitability of farmers.

With Tamil Nadu, the main crop has been rice as this state is blessed with two monsoons. But from 1992-1997, there has been a steady decline in the areas irrigated by canals and an increase in well-irrigated areas while the use of tanks remains an unreliable source of irrigation. However, major improvements in about 10 rice varieties released in the early 1990s can be expected to improve productivity growth in rice production although pests and diseases as well as imbalance in the use of fertilizers are major constraints. Thus Tamil Nadu could do with subsidies of pesticides and farmers should be educated on the more effective use of fertilizers to obtain high yields. Interestingly, the cropping pattern of late has shown increasing substitution of food crops by commercial crops but there is concern that the benefits will reach farmers only with the development of adequate infrastructure such as roads and markets. However, that Tamil Nadu has a higher index than the all India average of infrastructure.

Challenges of globalization

It is important to realize that globalization poses many challenges to a developing country like India, which had relied on a state directed and regulated policy regime for more than four decades. In moving to a more open, market-based economy there are many transitional problems that the country has to manage. The Government must play a pro-active role in facilitating the globalization process so that the opportunity sets for the economic agents are widened and the adverse effects of globalization are minimized. The Indian Government must also prepare the necessary information base and develop its capacity to articulate India's concerns and policy trade-offs in the international forums for multilateral trade and environmental negotiations.

In addition, the Government should embark on an extensive programme to educate farmers on the need to meet the standards required in the export markets. In fact, India needs to seek technical assistance in creating the capacity for meeting such standards and to consider watershed developments for environmental considerations. Equally important is the need to disseminate information about possible export markets to farmers, so that market access is achieved at minimum cost. Given the requisite information about markets and profitability, the likelihood of farmers investing in post-harvest and processing technologies and storage and efficient transportation arrangements as well as developing supporting infrastructure is very high.

Although the brave and bold move by India to reduce the tariff rate for agricultural products from 113 per cent in 1990-1991 to 26 per cent in 1997-1998 deserves to be applauded, the question of whether India is ready to compete in world markets remains to be seen. The infant industry argument may still hold for India to shield itself from external competition but one can easily question the length of time that is required to that end. Also, a delay in opening up to foreign trade has the danger that local producers may become too complacent and never be ready for competition.

As India opens up externally, it is also expected to face vulnerability in the wider international price fluctuations and thus Acharya (1998) claims that a minimum price support scheme is important. These prices can also act as a signal to adopt modern inputs and invest in yield-raising infrastructure for increasing production. For instance, keeping basic staple food grains at reasonable prices would induce farmers to switch over to high value crops. However, during the 1990s, Kalirajan and others (2001) shows that procurement prices especially for rice and wheat have been increasing faster than the general price level. Such high prices along with guaranteed purchases by the Food Corporation of India have pushed up market prices. These higher prices are partly responsible for the large buffer stocks with the Food Corporation. If this trend continues, India's comparative advantage will be eroded.

With openness and high price instability, unstable export revenue can also be expected. One way of reducing such risk is for India to diversify her agricultural exports. For example, since 1990, even in commodities such as tea, coffee, cocoa and spices, where India is supposed to have a comparative advantage (Chadha, 1999) international prices have been unstable. Besides increasing the type of exports to obtain more export revenue, India should also seriously consider exporting more value added agricultural products through agro-processing such as processed vegetables, fruits, fish and meat products given that export or even local demand for basic agricultural products would decline as incomes rise. The move to higher value added activities within the agricultural sector also spells greater opportunities for industrialization and vice versa as borne by Kalirajan and Shand's (1997) findings of a bi-directional relationship between agriculture and industry for most Indian states. On the other hand, Sivakumar and others (1999) establish empirical evidence of high forward linkages of agriculture due to the presence of agro-industries while Satyasai and Viswanathan (1999) show the significance of the spillover effects to the industrial sector via the intensive use of purchased inputs in the agricultural sector. The lack or slow pace of internal or domestic liberalization is also seen to hinder the possible gains from external or trade liberalization. For example, although central zoning restrictions have been abolished, state government restrictions on inter-state and even inter-district restrictions on marketing and movement of goods still exist in many cases. This interferes with the benefits from crop specialization and economies of scale arising from comparative advantage.

The land market is another example of distortion whereby land ceilings exist preventing the operation of large-sized farms. This has led to the emergence of a large number of small economically unviable land holdings. The easy leasing of land should be permitted with assurance of resumption. Yet another problem lies with the insufficiency of credit to agriculture. From 1995-1996, the Rural Infrastructure Development Fund was set up to allocate funds for the completion of projects and the government has committed itself to strengthening the cooperative credit structure through substantial refinancing and restructuring of the Regional Rural Banks. However, as mentioned earlier, due to varying institutional factors in the Indian states, these domestic reforms can be expected to yield quite different results.

Bibliography

A C Mittal: *Rural Economy*, Vista International, Delhi, 2006.

A Jayakumar: *Food Security Management : A Comparative Study*, Anmol Publication, Delhi, 2007.

A. Satyanarayana: *Andhra Peasants Under British Rule : Agrarian Relations and the Rural Economy 1900-1940*, Manohar Publication, Delhi, 1990.

Anshu: *Food Security in India : Challenges of 21st Century*, Kalakriti Offset, Delhi, 2009.

Archana Ruhela: *Agriculture and Food Security*, Oxford Book Company, Delhi, 2008.

Bimal N. Patel and Ranita Nagar: *Food Security Law : Interdisciplinary Perspectives*, Eastern Book Company, Delhi, 2014.

Biswajit Chatterjee and Asim K. Karmakar: *Food Security in India : A Comprehensive Plan for Sustainable Solution*, Regal Publications, Delhi, 2012.

Chaudhar. S.K.: *Rural Economy*, Sumit Enterprises, Delhi, 2009.

Dalip Kumar and Asmi Raza: *Agriculture and Food Security : Contemporary Issues*, Deep and Deep Publication, Delhi, 2011.

J.P. Arya: *Indian Rural Economy*, DND Publications, Delhi, 2015.

Jan Breman: *The Poverty Regime in Village India : Half a Century of Work and Life at the Bottom of the Rural Economy in South Gujarat*, Oxford University Press, Delhi, 2007.

Jitendra L Garmora: *Rural Development and Rural Economy*, Vital Publication, Delhi, 2007.

K.K. Tiwari: *Rural Planning and Development : Important Aspect of Indian Rural Economy*, DND Publications, Delhi, 2012.

Manjari Bhowmik Biswas: *Gender-Wise Distribution of Occupations : A Planning For Rural Economy*, Abhijeet Publication, Delhi, 2009.

Manob K. Bandyopadhyay: *Ganga-Farakka Barrage Project and Rural Economy of India*, Uppal Publication, Delhi, 2012.

Mustafa, A.: *Indian Rural Economy*, Serials Publication, Delhi, 2009.

Narayan Singh Rao: *Rural Economy and Society : Study of South-Eastern Rajasthan During the Eighteenth Century*, Rawat Publication, Delhi, 2002.

Niraj Kumar Verma: *Rural Economy of India : Globalization High Growth Trajectory Strategy for Inclusive and Holistic Development*, Deep and Deep Publication, Delhi, 2012.

P. Venkateshwara Rao: *Dairy Farm Business Management*, Biotech Books, Delhi, 2008.

Parmeshwar Singh: *Encyclopaedia of Farm Business Management*, Anmol Publication, Delhi, 2010.

Pradeep Chaturvedi: *Food Security in South Asia*, Concept Publication, Delhi, 2002.

Prem Nath: *Agriculture and Food Technology in Human Life*, Scientific Publication, Delhi, 2015.

R C Misro: *Rural Economy and Society of Ancient Orissa (C. A.D. 400-1000)*, Punthi Pustak, Delhi, 2002.

R. Patrick Wright: *The Standard Encyclopaedia of Modern Agriculture and Rural Economy (12 Vols-Set)*, Cosmo Publication, Delhi, 2006.

Ramniwas Sharma: *Agriculture and Food Safety*, Biotech Books, Delhi, 2006.

Shyam Kartik Mishra and Babita Agrawal: *Food Security in India : Policies and Challenges*, New Century Publications, Delhi, 2013.

Sudhansu Kumar Das and Sanjay Kanti Das: *Micro Finance and India's Rural Economy*, New Century Publications, Delhi, 2011.

Uma Narula: *Dynamics of Indian Rural Economy : Growth Perspective*, Atlantic Publication, Delhi, 2010.

Index